Numerical Analysis I
Lecture Slide Notes

Ralph E. Morganstern
Santa Clara University

ISBN-13: 978-1500946623
ISBN-10: 1500946621

Table of Contents

Preface _____ *viii*

1 *Introductory Material* _____ *1-1*

1.1 **Nature of Numerical Analysis** _____ **1-2**

1.2 **Numerical Algorithms** _____ **1-3**
1.2.1 Algorithms Behaving Badly _____ 1-4
1.2.2 Numerical Derivative of e^x is Zero _____ 1-5
1.2.3 MatLab Derivative Demo (5 sd) _____ 1-6
1.2.4 MatLab Derivative (2,5,8,10 sd) _____ 1-7
1.2.5 Polynomial Cancellation Effects _____ 1-8
1.2.6 Polynomial Cancellation Effects-2 _____ 1-9
1.2.7 Mean & Standard Deviation _____ 1-10
1.2.8 Standard Deviation Calculations _____ 1-11

1.3 **Scientific Computing** _____ **1-12**
1.3.1 Scientific Computing Environment _____ 1-13
1.3.2 Floating Point Representation _____ 1-14
1.3.3 Example: Base 10, 2 digits, 4 decades _____ 1-15
1.3.4 Typical Floating Point Systems _____ 1-16

1.4 **Calculus Concepts for Numerical Analysis** _____ **1-17**
1.4.1 Mapping, Limit, Continuity _____ 1-18
1.4.2 Sequences & Limits _____ 1-19
1.4.3 Differentiable Functions & Continuity _____ 1-20
1.4.4 Theorems Relating to Error Estimation _____ 1-21
1.4.5 Weighted Mean Value / Intermediate Value _____ 1-22
1.4.6 Generalized Rolle & Taylor _____ 1-23

1.5 **Error Analysis** _____ **1-24**
1.5.1 Truncation & Round-off Error _____ 1-25
1.5.2 Derivation of Truncation & RO Errors _____ 1-26
1.5.3 Step Size Dependence of Total Error _____ 1-27

1.6 **Mathematical (Inherent) Instability** _____ **1-28**
1.6.1 Sensitivity & Condition Number-Details _____ 1-29

1.7 **Numerical Instability** _____ **1-30**
1.7.1 Solution to a Difference Equation _____ 1-31
1.7.2 Example of Numerical Instability _____ 1-32

1.8 **Reducing Round-Off Error** _____ **1-33**
1.8.1 New Method of Computation -1 _____ 1-34
1.8.2 New Method of Computation -2 _____ 1-35
1.8.3 New Method of Computation -3 _____ 1-36
1.8.4 Double Precision Subtraction _____ 1-37

2 *Root Finding Techniques* _____ *2-1*

2.1 **Iterations & Stopping Criteria** _____ **2-2**

Table of Contents

2.2   **Root Finding Overview** _____ **2-3**

2.3   **Bracketing Methods - Bisection** _____ **2-4**
    2.3.1   Bracketing Methods – False Position _____ 2-5

2.4   **Non-Bracketing – Secant** _____ **2-6**
    2.4.1   Extension of Secant – Muller's Method _____ 2-7
    2.4.2   Muller's Method Details _____ 2-8
    2.4.3   Non-Bracketing – Newton-Raphson _____ 2-9
    2.4.4   Problems With Newton-Raphson _____ 2-10

2.5   **Root Multiplicity & Convergence Rate** _____ **2-11**

2.6   **Extensions of Newton-Raphson** _____ **2-12**
    2.6.1   Extensions of Newton-Raphson -2 dim _____ 2-13

2.7   **Convergence Rates Comparison** _____ **2-14**

2.8   **Fixed Point Methods** _____ **2-15**
    2.8.1   Fixed Point Sequence _____ 2-16
    2.8.2   Fixed Point Theorem & Error Bounds _____ 2-17
    2.8.3   Fixed Point Trajectories _____ 2-18
    2.8.4   Fixed Point Iterates Example _____ 2-19

2.9   **Root Finding Methods Compared: 5 sd Good x_0** ____ **2-20**
    2.9.1   Root Finding Methods Compared: 5 sd Bad x_0 _____ 2-21
    2.9.2   Non-Bracketing Root Finding Methods: 16 sd Good x_0 ____ 2-22
    2.9.3   Non-Bracketing Root Finding Methods: 16 sd Bad x_0 _____ 2-23

2.10   **Polynomial Roots** _____ **2-24**
    2.10.1   Horner's Method - Recursions _____ 2-25
    2.10.2   Horner's Method - Tableau _____ 2-26
    2.10.3   Horner's Method - Example _____ 2-27
    2.10.4   Computational Efficiency of Horner's Method _____ 2-28
    2.10.5   Polynomial Root Finding Methodology _____ 2-29

2.11   **Accelerating Convergence of Iterates** _____ **2-30**
    2.11.1   Equivalence of $\Delta p_n / \Delta p_{n-1}$ to $\varepsilon_n / \varepsilon_{n-1}$ _____ 2-31
    2.11.2   Aitken's Method: $g(x) = \ln(2\cos x)$ _____ 2-32
    2.11.3   Aitken's Acceleration (MatLab) _____ 2-33
    2.11.4   Aitken's Method – Math Details _____ 2-34

3   _**Polynomial Interpolation**_ _____ _**3-1**_

3.1   **Fitting a Curve to Tabulated Values** _____ **3-2**

3.2   **Polynomial Properties** _____ **3-3**
    3.2.1   Why Different Polynomials? _____ 3-4
    3.2.2   Polynomial Overview - 1 _____ 3-5
    3.2.3   Vandermonde, Taylor, Lagrange, Hermite _____ 3-6
    3.2.4   Polynomial Overview - 2 _____ 3-7
    3.2.5   Hermite, Laguerre, Tschebyshev, Legendre _____ 3-8

Table of Contents

3.3   Taylor Polynomials _____ **3-9**
 3.3.1   Taylor Polynomials - Example _____ 3-10

3.4   Lagrange Interpolating Polynomials _____ **3-11**
 3.4.1   Lagrange Interpolating Polynomials - General _____ 3-12
 3.4.2   Lagrange Poly: Effects of Adding Points _____ 3-13
 3.4.3   Example: Linear Combination of Basis Polynomials _____ 3-14
 3.4.4   Creating Table With Specified Error _____ 3-15
 3.4.5   Lagrange Interpolation Table - Details _____ 3-16
 3.4.6   Choosing Degree of Lagrange Polynomial _____ 3-17

3.5   Iterated Interpolation Tableau _____ **3-18**
 3.5.1   Iterated Interpolation for Bessel Function _____ 3-19

3.6   Newton Divided Difference Polynomials _____ **3-20**
 3.6.1   Divided Difference Table _____ 3-21
 3.6.2   Forward, Backward, and Central Polynomials _____ 3-22
 3.6.3   Inverse Newton Interpolation Polynomials _____ 3-23

3.7   Uniformly Spaced Difference Tables _____ **3-24**
 3.7.1   Physics Lab Spark Strip & "g" _____ 3-25

3.8   Hermite Interpolating Polynomials _____ **3-26**
 3.8.1   Hermite Poly Example _____ 3-27

3.9   Cubic Splines _____ **3-28**
 3.9.1   Cubic Spline Algorithm _____ 3-29
 3.9.2   Free Cubic Spline Example _____ 3-30

3.10   Tschebyshev Polynomials and Economization _____ **3-31**
 3.10.1   Tschebyshev Economization _____ 3-32

3.11   Bezier Parametric Curves & Computer Graphics _____ **3-33**
 3.11.1   Bezier Guide Points & Scale Factors _____ 3-34
 3.11.2   MatLab Bezier Curves _____ 3-35
 3.11.3   MatLab Bezier Curves-2 _____ 3-36
 3.11.4   MatLab Bezier Curves-3 _____ 3-37

4   Derivatives and Integrals _____ **_4-1_**

4.1   Computation using Polynomial Interpolates _____ **4-2**
 4.1.1   Derivative of Lagrange Polynomial _____ 4-3
 4.1.2   Derivative Formulas _____ 4-4
 4.1.3   Derivative Formulas Application _____ 4-5
 4.1.4   Differentiation – Taylor Tricks _____ 4-6

4.2   Instability of Numerical Derivatives _____ **4-7**

4.3   Richardson Extrapolation _____ **4-8**
 4.3.1   Mathematical Details _____ 4-9
 4.3.2   Extrapolate from 3-point to 5-point Derivative Formula __ 4-10

4.4   Extrapolation Table Stopping Criteria _____ **4-11**

Table of Contents

    4.4.1   Extrapolation Tables 5sd & 8sd _____ 4-12

4.5   Integral of Lagrange Polynomial _____ **4-13**
    4.5.1   Integration Formulas (Quadrature) _____ 4-14
    4.5.2   Composite Methods _____ 4-15
    4.5.3   Composite Trapezoid & Simpson _____ 4-16

4.6   Stability of Composite Quadrature Methods _____ **4-17**

4.7   Adaptive Quadrature _____ **4-18**
    4.7.1   Adaptive Quadrature Algorithm Flow _____ 4-19
    4.7.2   Adaptive Quadrature Algorithm Implementation Issues_____ 4-20
    4.7.3   Adaptive Quadrature Examples _____ 4-21

4.8   Romberg Integration_____ **4-22**
    4.8.1   Romberg Integration-Examples -1 _____ 4-23
    4.8.2   Romberg Integration-Examples -2 _____ 4-24

4.9   Gaussian Quadrature _____ **4-25**
    4.9.1   2-pt Gaussian Quadrature _____ 4-26
    4.9.2   Gaussian Quadrature – Intuitive Proof _____ 4-27
    4.9.3   n-pt Gaussian Quadrature _____ 4-28
    4.9.4   Other Orthonormal Quadratures _____ 4-29
    4.9.5   Orthonormal Quadratures - Example _____ 4-30
    4.9.6   Multi-panel Gauss Quadratures _____ 4-31
    4.9.7   Multi-Panel Gauss Quadrature Examples _____ 4-32
    4.9.8   Integration Methods Comparison _____ 4-33

4.10 Improper Integral Techniques _____ **4-34**
    4.10.1   Improper Integrals –Examples _____ 4-35
    4.10.2   Improper Integrals with End Point Singularities _____ 4-36
    4.10.3   Simpson Estimate of Improper Integral _____ 4-37

4.11 Multiple Integrals _____ **4-38**
    4.11.1   Simpson Single 2^d Cell Weight Matrix _____ 4-39
    4.11.2   Composite Simpson Multi-Cell Weight Matrix 2^d _____ 4-40
    4.11.3   Composite 4-Cell Simpson Quadrature Example _____ 4-41
    4.11.4   Simpson Non-Rectangular Cell – 2^d _____ 4-42
    4.11.5   Simpson Non-Rectangular Cell – 3^d _____ 4-43
    4.11.6   Gauss Cell Coefficient Matrix Pattern – 2^d _____ 4-44
    4.11.7   Gaussian Non-Rectangular Cells – 3^d _____ 4-45
    4.11.8   Simpson Gauss Quadratures Comparison - Cone Example _____ 4-46
    4.11.9   Multiple Integrals Methods Comparison_____ 4-47

5   _MatLab Scripts_ _____ **_5-1_**

5.1   Derivative of exp(x) _____ **5-2**

5.2   Polynomial Cancellation _____ **5-3**

5.3   Fixed Point Example _____ **5-4**

5.4   Aitken's Convergence Acceleration _____ **5-5**

Table of Contents

5.5   **Taylor Polynomials for 1/x**   5-6

5.6   **Lagrange Polynomials for Sin(x) and Asin(x)**   5-7

5.7   **Richardson Extrapolation**   5-10

5.8   **Bezier Curves**   5-11

5.9   **Bezier Curves - Interactive**   5-12

5.10   **Root Finding Methods Comparison**   5-13

5.11   **Simpson 1^d Adaptive Quadrature**   5-16

5.12   **Gauss 1^d Multi-Panel Quadrature**   5-21

5.13   **Simpson 1^d Composite with fixed intervals**   5-28

5.14   **Quadratures – Simple 2^d Simpson & Gaussian**   5-29

5.15   **General 2^d Simpson Quadrature & Weight Matrix**   5-30

5.16   **General 3^d Simpson Quadrature & Weight Matrix**   5-33

5.17   **General 2^d Gaussian Quadrature**   5-37

5.18   **General 3^d Gaussian Quadrature**   5-40

6   *References*   *6-1*

7   *Index*   *7-1*

Preface

These Lecture Slide Notes have been used over the past several years for a two-quarter graduate level sequence in numerical analysis. Part 1 covers introductory material on the Nature of Numerical Analysis, Root Finding Techniques, Polynomial Interpolation, Derivatives, and Integrals. Part 2 covers Ordinary Differential Equations and Numerical solutions to Linear Systems of Equations.

This "Lecture Slide Notes" format is convenient for self-study because it covers the subject matter in a concise and easily accessible manner by employing multiple visualization techniques in slide format together with focused explanatory notes. Each slide stands alone as a "one-page synopsis" that encapsulates a complete concept, algorithm, or theorem using a combination of equations, graphs, diagrams, plots, illustrative tableaus and comparison tables. The explanatory notes are placed directly below each slide in order to reinforce and/or give additional insight into the particular numerical technique or concept illustrated in the slide.

Part 1 starts with a discussion of the nature of numerical computation and scientific computing by considering the role of number systems (significant digits) and computer word size (precision) in producing a useful algorithm. The trade off between truncation and round off errors is introduced by considering some examples of algorithms behaving badly. Basic calculus theorems are reviewed and subsequently used to analyze error growth, and some simple techniques to reduce it. The basic root finding techniques, their interrelations, generalizations, and convergence rates are discussed and explicit example computations and comparison tables are displayed. Horner's tableau method for synthetic division is analyzed in detail together with its relation to the notion of numerically efficient computations with polynomials *via* nesting. The concept of polynomial fits through nodal points leads to interpolation and extrapolation concepts which are useful in their own right, but also feed directly into the evaluation of numerical derivatives and integrals. Lagrange, Taylor, Newton, Hermite, and Tschebyshev polynomials, as well as cubic splines and Bezier curves are all discussed here. Finally the use of polynomials to calculate derivatives and integrals are explored along with iterative and extrapolation methods used to improve their accuracy and usability. Methods discussed include Newton-Cotes, composite Simpson and trapezoidal, Romberg and Gaussian quadrature methods. Finally a brief discussion of improper and multiple integrals are given together with some explicit examples.

A Table of Contents serves to organize the slides in terms of the main numerical analysis topics covered and gives a complete list of slide titles and their page numbers. An index is also provided to link related aspects of topics and cross-reference key concepts, specific applications, and the various visualization aids. Although no problem sets have been included in these notes, a good number of examples are worked out in detail. Moreover, Section 5 contains a selection of illustrative MatLab scripts that are keyed to these examples and can easily be edited by the student to generate solution tables and plots for similar problems. Finally, references to a number of standard text books are given, but there has been no attempt to make an exhaustive bibliography.

1 Introductory Material

1.1 *Nature of Numerical Analysis*

Nature of Numerical Analysis

- **Numerical Algorithm is a Method of Calculation**
  - Limited Number of Digits (***Precision***)
  - Errors Accumulate (***Accuracy***)
  - Algorithm Efficiency (***Convergence Rate***)
- **Address Real Physical Phenomena**
  - Model Parameters Affect Nature of Solution (***Stability***)
  - Applicable to Wide Range of Parameters (***Robustness***)
  - Floating Pt Number Representation *(**Dynamic Range**)*
- **Scientific Computing Methodology**
  - Approximate Physical Model (***Parameter Range***)
  - Approximate Math (***Polynomial Approximation***)
  - Error Analysis (***Truncation, Rounding, Propagation***)

A numerical solution is more than just a stream of numbers that takes the place of an analytic solution to an engineering problem. The output is generated by some computational algorithm and even assuming that the algorithm is formulated and programmed correctly the stream of numbers can still be complete nonsense. A major part of numerical analysis is to provide a systematic understanding of the nature of numerical errors and instabilities and insure that the algorithm remains valid over a wide range of engineering parameters.

The number of digits in the computer word (precision), the numerical error (accuracy), the algorithm efficiency (convergence rate), its ability to sustain small parameter perturbations (stability), its applicability to a range of problems (robustness) and its invariance under large scale changes (dynamic range) are all important characteristics of the specific algorithm. The control of and trade-off between these characteristic algorithm properties constitutes the main thrust of Numerical Analysis.

Scientific Computing attempts to predict real world experimental observations by using a set of known (observed) physical parameters to develop a Physical Model based on mathematical approximations. The resulting computations of the Numerical Algorithm are analyzed and adjusted to make the accuracy level of the output data commensurate with those of the physical observations. Subsequent comparison with the observed data allows for corrections to the physical model and/or its mathematical approximations.

Numerical Algorithms

- Algorithms Behaving Badly
- Scientific Computing
- Floating Point Representation

In order to emphasize the fact that we should not simply accept a stream of numbers generated by some computational algorithm that has been formulated and programmed correctly, we look at some simple algorithms programmed in MatLab[©]. These "algorithms behaving badly" show that blind faith in such output streams leads to erroneous results and therefore sets the stage for a closer look at the art of Scientific Computing, its associated Floating Point Representation of numbers, and an analysis of all types of errors that "creep into" these bad algorithms.

1.2.1 Algorithms Behaving Badly

Algorithms Behaving Badly

- Numerical Derivative of e^x is Zero
- MatLab Derivative Demo (5 sd)
- Polynomial Cancellation Effects
- Mean & Standard Deviation

In the following slides we consider a few examples of "algorithms behaving badly."

We first take the standard calculus definition of a derivative as the limit of a sequence of functional differences divided by the stepsize "h" *viz.,* $f'(x) = \{f(x+h)-f(x)\}/h$ in the limit as $h \rightarrow 0$.

The MatLab$^\copyright$ script (see Appendix A) yields the surprising result that the stream of data approaches zero independent of how many significant digits (precision) we use; clearly this is not an acceptable result. Only a careful error analysis will give us a clear understanding of the optimal choice of the stepsize h for accurate numerical results.

In the second example, a high degree polynomial $(x-1)^6$ is evaluated directly and compared with the evaluation of its expanded form. Such a polynomial is very flat near the point x=1 and upon going to smaller and smaller scales about x=1 the MatLab$^\copyright$ script generates "random" looking data with large excursions from the accurate factored form of the polynomial. This is the result of cancellation effects in the expanded form of the polynomial because of the limited precision arithmetic. Some improvement can be effected by nesting the polynomial; but of course this is still not equivalent to using the fully factored form.

In the final example, the Standard Deviation, σ, of a simple set of data is computed by two different methods using limited precision arithmetic. It is shown that σ can be computed by performing either one or two passes through the data; the surprising fact is that the one-pass method gives extremely bad results, whereas the two-pass method yields full precision results.

All three examples serve notice that correct mathematics and solid programming are not sufficient to insure accurate results.

1.2.2 Numerical Derivative of e^x is Zero

Numerical Derivative of e^x is Zero

$$f'(x_0) = \frac{f(x_0 + \Delta x) - f(x_0)}{\Delta x} + \text{error}$$

$$f(x) = e^x \qquad \Delta x = h = 2^{-k} \; ; \; k = 1, 2, \cdots, 12$$
$$x_0 = 1$$

$$f'(x)\big|_{x=1} = e^x\big|_{x=1} = e = 2.718281828\cdots$$

$$f'(x) \cong D^{(k)} \equiv \frac{f(x_0 + 2^{-k}) - f(x_0)}{2^{-k}}$$

The numerical derivative of the exponential (or, in fact, any function) approaches zero in the "theoretical" calculus limit. We take the elementary definition of a derivative as the difference between the function evaluated at $x_0 + h$ and at x_0 divided by h and subsequently halve h at each step of the sequence. The exact calculus value of the derivative is given as 2.71828..., but as the next slide shows the limit of this numerical sequence gives zero for the derivative.

1.2.3 MatLab Derivative Demo (5 sd)

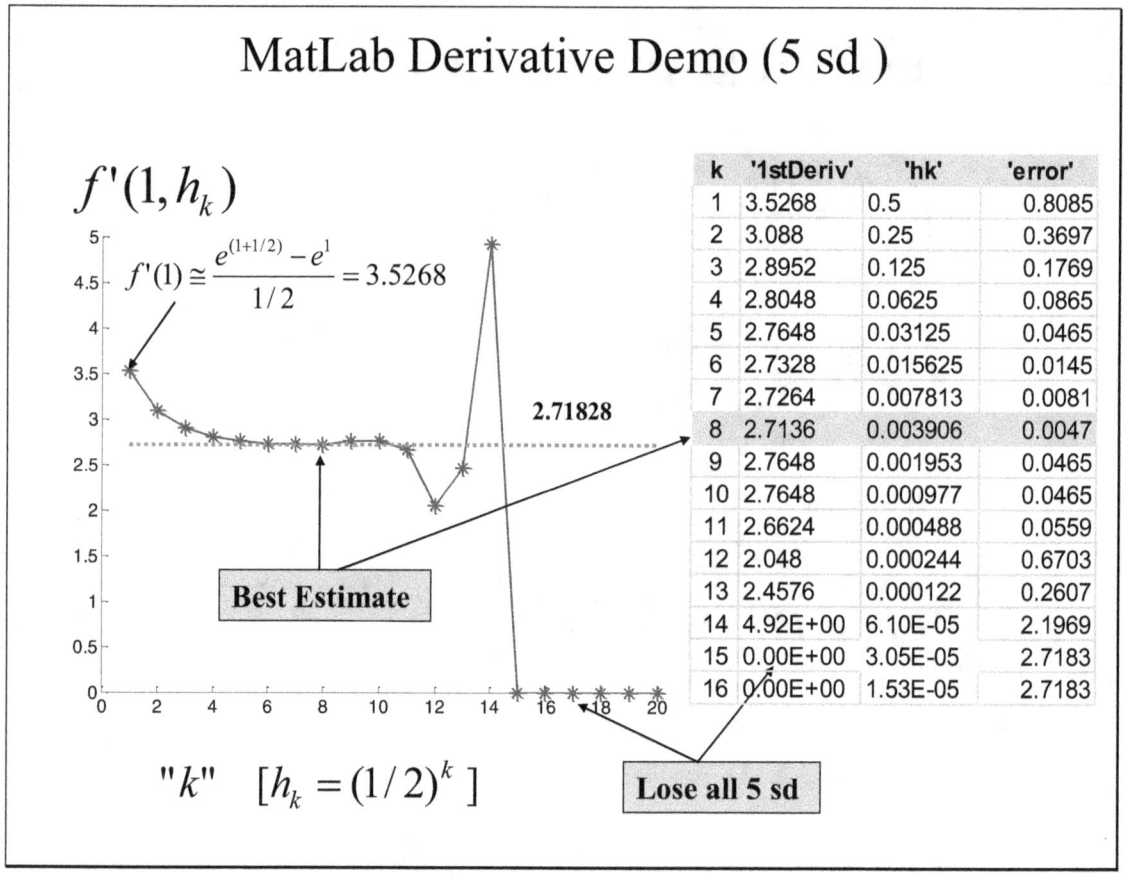

We employ 5 significant digit arithmetic and display the values of the derivative as a function of index k in the table and then graph the derivative as a function of the halving index k. It is seen that the value of the derivative is closest to the calculus value for k=8 where f'(1,k=8)=2.7136 and is pretty flat immediately before and after, but then becomes erratic and finally drops to zero and stays there! (See MatLab script on Slide#5-2)

1.2.4 MatLab Derivative (2,5,8,10 sd)

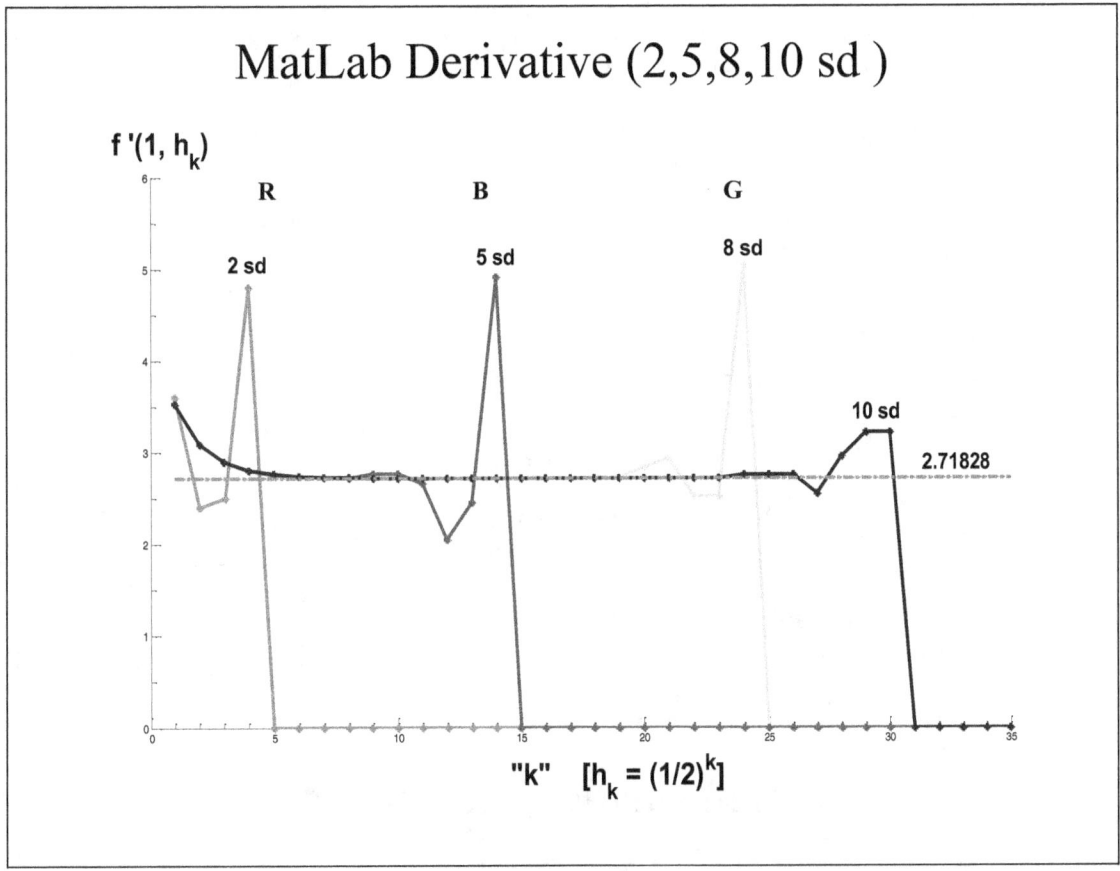

We graphically compare the results for 2, 5, 8, and 10 digit arithmetic and they all display similar behavior reaching a "best" value at some value of k remaining flat near this value of k and then becoming erratic and finally dropping to zero. Clearly increasing the precision of the calculation just delays, but does not prevent the eventual drop to zero. There is something fundamentally wrong with this algorithm because no matter how many significant digits we use, all precision is eventually lost and we wind up with the obviously incorrect result that the derivative of the exponential evaluated at x=1 is zero. (See MatLab script on Slide#5-2)

1.2.5 Polynomial Cancellation Effects

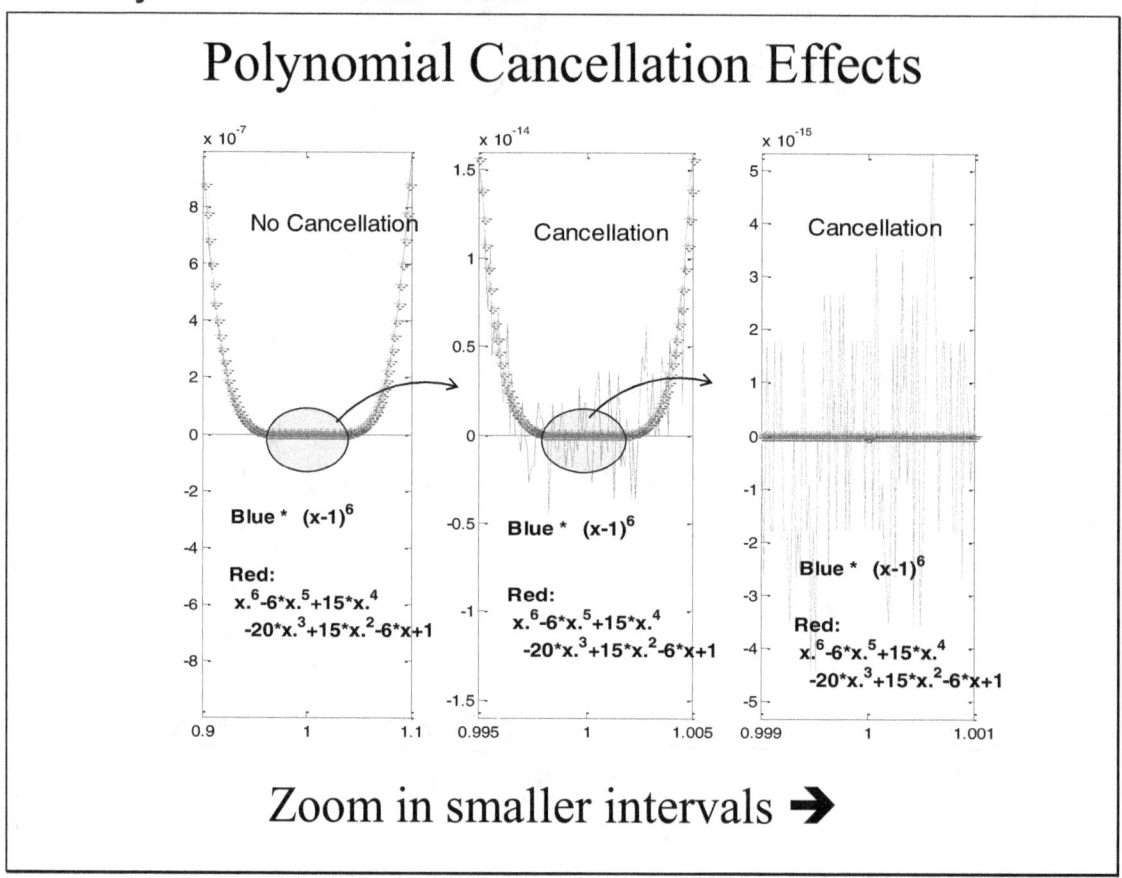

Evaluating a closed polynomial expression such as $(x-1)^6$ using a fully expanded polynomial also displays an anomaly. The sequence from left to right shows smaller and smaller intervals and we see that the polynomial evaluation (red) starts to diverge from the true value (obtained from the original polynomial) and in fact appears quite random. This results from cancellation effects between the various terms in the expanded polynomial for the limited precision arithmetic. (See MatLab script on Slide#5-3)

1.2.6 Polynomial Cancellation Effects-2

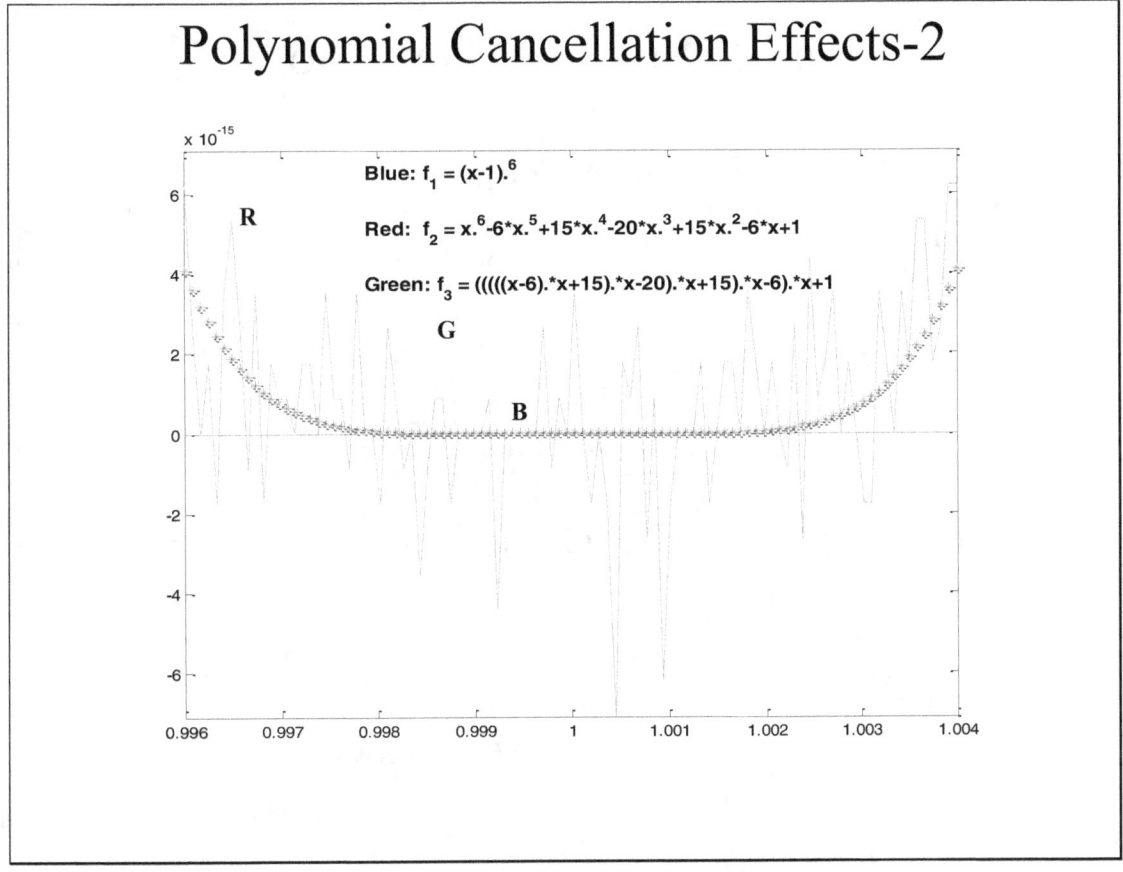

Here we see that nesting the polynomial (green) helps somewhat but still appears random compared to the original polynomial. We will see that many numerical algorithms employ polynomial expansions as approximations to analytic functions and hence are subject to these cancellation effects; the nesting of polynomials before evaluation is one technique to reduce such errors. (See MatLab script on Slide#5-3)

1.2.7 Mean & Standard Deviation

Mean & Standard Deviation

- ## Two Passes Through Data

$$\bar{x} = \frac{1}{N} \sum_{k=1}^{N} x_k \qquad \text{Mean: Pass\#1}$$

$$\sigma = \sqrt{\frac{1}{N-1} \sum_{k=1}^{N} (x_k - \bar{x})^2} \qquad \text{Sdev: Pass\#2}$$

Add small #'s

- ## Single Pass for Std Deviation

$$\sigma = \sqrt{\frac{1}{N-1} \left(\sum_{k=1}^{N} x_k^2 - N \cdot \bar{x}^2 \right)}$$

$$= \sqrt{\frac{1}{N-1} \left(\sum_{k=1}^{N} x_k^2 - \frac{1}{N} \left(\sum_{k=1}^{N} x_k \right)^2 \right)}$$

Subtract nearly equal #'s

Loss of Significant Digits

Two methods for computing the standard deviation of a data set are given.

Two Passes: This requires two passes through the data one to compute the mean and the second to compute the standard deviation from the mean. It may be thought at first that this is an inefficient method; but it turns out to be a better numerical method than 2)

Single Pass: Substituting the definition of the mean into the standard deviation expression allows us to reduce it to the second form which only requires a single pass through the data. This is a more efficient algorithm, but is a very bad algorithm from a numerical standpoint.

1.2.8 Standard Deviation Calculations

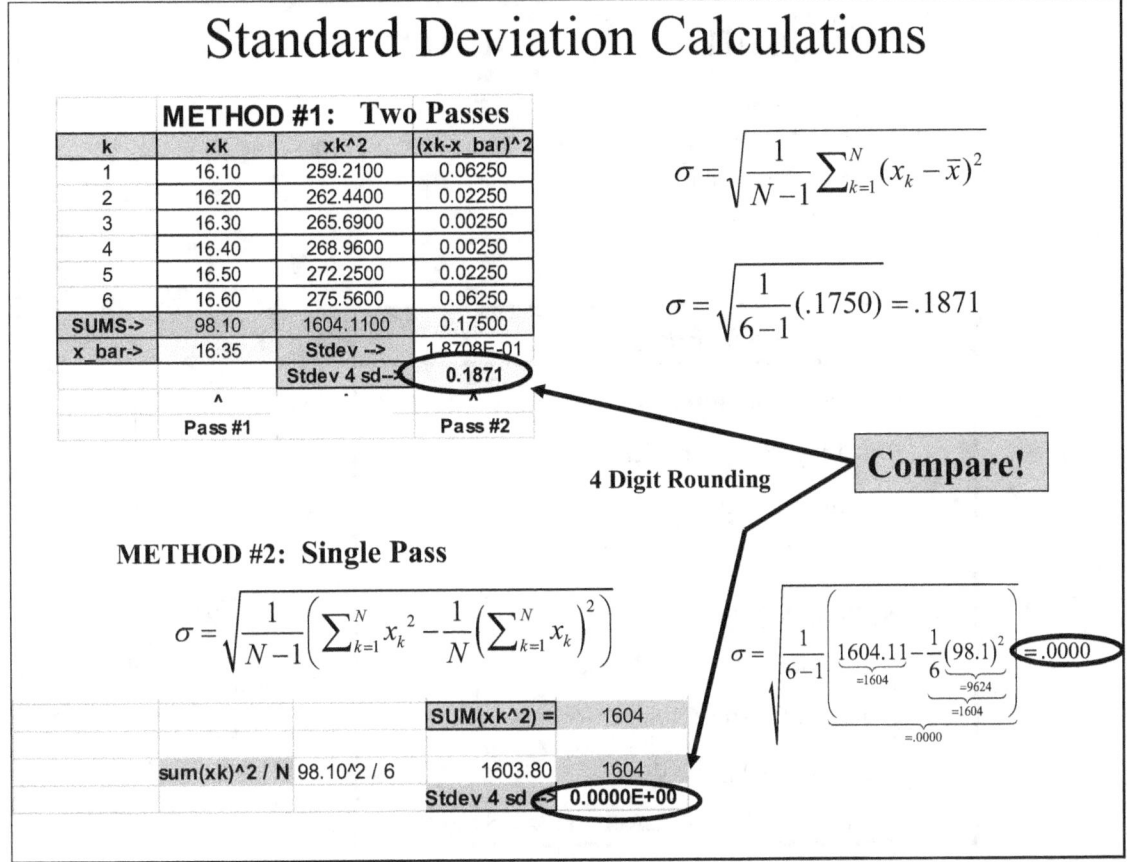

Here is a simple standard deviation computation example with six data values having 4 significant digits {16.10, 16.20,16.30,16.40,16.50,16.60}.

In the 2-pass procedure, the mean is computed to be 16.35 and then the standard deviation is computed in the table to be 0.1871.

In the 1-pass procedure, the standard deviation is computed directly from the formula and all 4 significant digits are lost in the process yielding a standard deviation of 0.0000

1.3 *Scientific Computing*

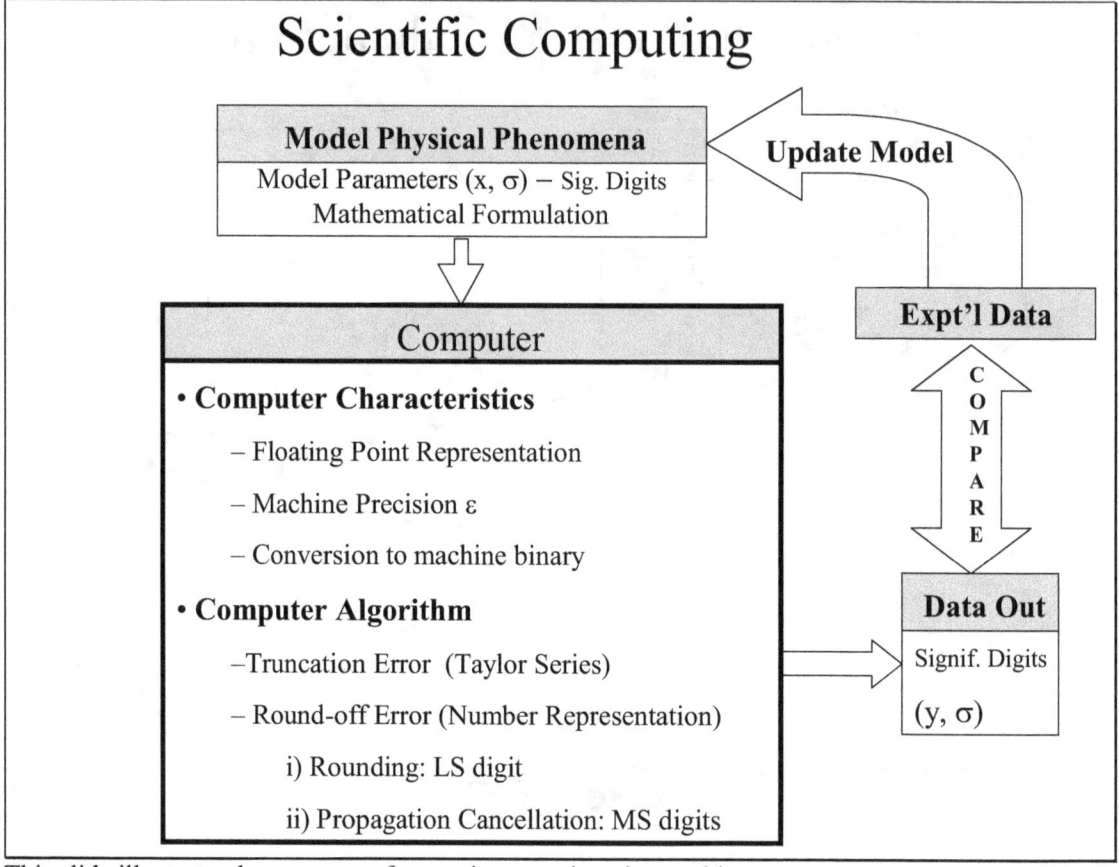

This slide illustrates the anatomy of errors in an engineering problem.

Modeling Error: A model is formulated mathematically and relies on some physical parameters that are known with a given accuracy.

Algorithm Error: The model is then formulated as a computer algorithm which always requires some mathematical approximations (truncation error) and some manipulation in the way actual numbers are represented and their products and sums are evaluated (round-off and propagation errors).

Number Conversion Error: The computer operates on binary numbers so that there are also errors that occur in the required input and output conversions from decimal to/from binary.

Experimental Accuracy Error: Experimental error in testing a theory against the predicted computer output yields new values for the fundamental parameters which are then used to update the physical model.

Given that the numerical algorithm has been carefully constructed to represent the physical model with an accuracy equivalent to the experimental observations, any difference between the predicted values and those actually observed must be attributed to incorrect physical modeling. These differences must then be accounted for by a new physical model. This is iterative procedure in which more accurate measurement techniques require new physical models which in turn require new numerical algorithms and higher precision computers. Only idealized problems can be solved analytically and inevitably we must resort to numerical computations for real world problems.

1.3.1 Scientific Computing Environment

Scientific Computing Environment

	Characteristic	Effect
Experimental Validation	1) Valid Math Model	Accurate Results
	2) Noisy data is inaccurate	Independent of precision (#digits)
	3) High precision computer	Utilizes full data accuracy
Algorithm Errors	4) Truncation & RO errors	Depend oppositely on step size h
	5) Number Representation errors	Data input/output
	6) Round off errors (rounding)	Affect Least SD
	7) Cancellation errors (propagation)	Affect Most SD
	8) Singularity-free Equations	Avoid inherent instabilities
Algorithm Stability	9) Stable Algorithms	Avoid numerical instabilities
	10) Algorithms are Recipes	Limit error growth
	11) Practical Trade offs necessary	Tailored Numerical Solutions

The implementation of an engineering model using a numerical algorithm has elements that relate to i) experimental validation ii) algorithm errors, and iii) algorithm stability as illustrated in the table.

Experimental Validation (*Rows# 1-3*): A *valid math model* predicts results that match experimental observations with the *desired accuracy*. However, we should avoid precision overkill. We should match the significant digits in the algorithm to the expected accuracy of the experimental results

Algorithm Errors (*Rows# 4-8*): The mathematical truncation errors decrease with stepsize h while numerical round-off errors increase with stepsize. Thus, there is an optimum stepsize that minimizes the total error. Round-off errors caused by input/output conversions and algebraic propagation affect the least significant digits, while cancellation errors (as in polynomial expansions) affect the most significant digits.

Algorithm Stability (*Rows# 9-11*): Model equations must not have singularities; numerical algorithms have their own stability issues. Error growth cannot be eliminated, but it can be controlled by tailoring the algorithm to a specific class of engineering problems

1.3.2 Floating Point Representation

Floating Point Representation

- Base β, mantissa-digits n, exponent $\{\beta, n, L, U\}$

$$e \in [L, U]$$

$$\hat{x} \equiv fl(x) = \pm 0.\underbrace{d_1 d_2 d_3 \cdots d_n}_{mantissa} \times \underbrace{\beta}_{Base}^{\overset{Exp}{e}} \qquad d_k = \begin{cases} 1, \cdots \beta - 1 & k = 1 \\ 0, 1, \cdots \beta - 1 & k \neq 1 \end{cases}$$

- Machine uses floating point binary digits $\beta = 2$
- User input/output in floating point decimal $\beta = 10$
- Relative Error ("Precision"," Machine Epsilon")

$$\varepsilon_{rel} \equiv \frac{|fl(x) - x|}{|x|} \leq \frac{\beta}{2} \cdot \beta^{-n} \quad \xrightarrow[\text{base 10}]{} \quad \varepsilon_{rel} \leq 5 \cdot 10^{-n}$$

A computer word consists of a fixed number of binary digits and when left in strict binary form the dynamic range of numbers represented from smallest to largest can be quite inadequate for scientific problems.

In order to represent numbers over a large dynamic range floating point representations are used. The floating point number has four components, namely a base, the number of mantissa digits, and an upper and lower exponent to express the dynamic range.

Floating point numbers start with a sign followed by a decimal point, a non-zero first digit, and then the remaining unrestricted digits.. There are n digits in the mantissa and the base is raised to an exponent in the allowed range.

The precision or "machine epsilon" is the worst relative error caused by the floating point number representation. It is equal to the **maximum** absolute deviation between the true number x and its floating point representation divided by x and is given by the formula for relative error on the slide.

1.3.3 Example: Base 10, 2 digits, 4 decades

Example: Base 10, 2 digits, 4 decades

$$\{\beta, n, L, U\} = \{10, 2, -1, 2\} \qquad \Rightarrow x = \underbrace{\pm}_{2} . \underbrace{d_1}_{9} \underbrace{d_2}_{10} \times \underbrace{10^e}_{4}$$

- Finite: 720 Numbers $\qquad x \in [.10 \times 10^{-1}, .99 \times 10^{+2}\}$

- Not Uniformly Spaced ("Log10 Uniform")

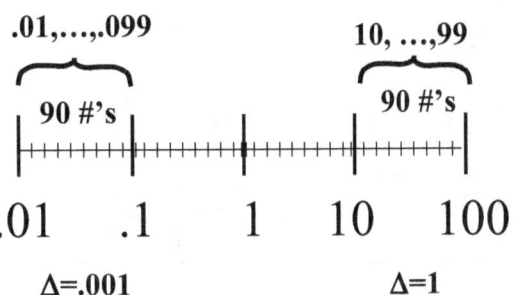

This simple base 10 example uses two decimal digits and 4 decades and yields 720 numbers spread over a dynamic range of 4 decades from .01 to 100. Note that there are 90 distinct numbers in each decade resulting in 360 numbers for each sign. These numbers are not uniformly spaced; rather they are logarithmically uniform. That is, in each of the four decades [.01 to .1), [.1 to 1), [1 to 10), [10 to 100) there are exactly 90 numbers that are uniformly distributed across the decade. The separation in the first decade is $90 * \Delta = .1 - .01 \rightarrow \Delta = .001$, the second is $90 * \Delta = 1 - .1 \rightarrow \Delta = .01$, the third is $90 * \Delta = 10 - 1 \rightarrow \Delta = .1$, and finally, the fourth is $90 * \Delta = 100 - 10 \rightarrow \Delta = 1$. Thus, the mantissas are uniformly spaced but they are multiplied by different powers of 10.

1.3.4 Typical Floating Point Systems

Typical Floating Point Systems

System	Base β	Precision n	Lower Expon. L	Upper Expon. U
IEEE SP	2	24	-126	127
IEEE DP	2	53	-1022	1023
Cray	2	48	-16,383	16,384
HP Calc	10	12	-499	499
IBM MainF	16	6	-64	63

This table (adapted from Ref. 1 p. 18) of typical floating point systems shows the tradeoff between mantissa precision and dynamic range; that is how you allocate between mantissa and exponent.

Note that going from IEEE Single to Double Precision (DP) not only increases the precision of the mantissa but also expands the dynamic range by a factor of ~10. The Cray has a little less precision than IEEE DP, but increases the dynamic range by a factor of ~16 giving it a huge dynamic range!

The HP Calculator has less precision than IEEE SP but has an impressive dynamic range. By comparison, the IBM Mainframe (which essentially put us on the Moon) has ½ the precision and a ~1/10 the dynamic range of the HP 48 series Calculator.

Calculus Concepts for Numerical Analysis

- Mapping, Limit, Continuity
- Sequences & Limits
- Differentiable Functions & Continuity
- Theorems Relating to Error Estimation
    - Rolle Theorem (*Two Zeros and "Turning"*)
    - Mean Value Theorem (*Secant Slope -Tangent*)
    - Extreme Value Theorem (*Min & Max*)
    - Weighted Mean Value Theorem (*Weighted Averaging*)
    - Intermediate Value Theorem (*Zero Crossing & Root Finding*)
    - Generalized Rolle Theorem (*Multiple Zeros*)
    - Taylor Theorem (*Analytic Approximation & Error Estimate*)

This is a review of some basic calculus concepts and theorems which are used in numerical analysis. It is not the main thrust of the course, but you need to understand these theorems in order to follow the derivations and to really get a feeling for the analysis which supports many of the numerical methods and restrictions to their validity. Note that the applicability of each theorem is summarized by key words in the parentheses.

1.4.1 Mapping, Limit, Continuity

Mapping, Limit, Continuity

- Function Mapping f: X→Y X,Y ε R
- Accumulation Point or Limit

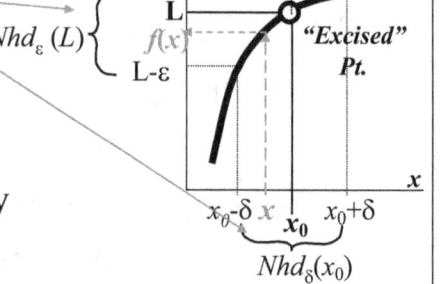

If for every $\varepsilon - Nhd(L)$

there corresponds a $\delta - Nhd(x_0)$: $|x - x_0| < \delta$

such that $|f(x) - L| < \varepsilon$ for all $x \in Nhd_\delta(x_0)$

say limit exists: $\lim_{x \to x_0} f(x) = L$

- **Note:** Limit pt. **"L"** is not necessarily
  the same as $f(x_0)$;
- Rather L is at the place where $f(x_0)$
  "should be" had we not excised it

- Continuous $f(x)$ at x_0 means: $f(x_0)=L$

A function is *any mapping* of the set X to the set Y=f[X] and usually designated by f: X→Y.
An Invertible function requires a *1-to-1 and onto mapping* ("invertible mapping") . This means explicitly that "f maps each x onto a unique y, so that "f-inverse" f^{-1} pulls each y back to the unique x from which it came , *i.e.*, x = f^{-1} (y) for all x and y.
L is an Accumulation Point (limit point) of function y = f(x)
if for every small ε-neighborhood about the point L, $Nhd_\varepsilon(L)$=[L-ε, L+ε], there exists a corresponding small δ-neighborhood about the point x_0, $Nhd_\delta(x_0)$=[x_0 -δ, x_0 +δ], such that all points x in $Nhd_\delta(x_0)$ map back to points y in $Nhd_\varepsilon(L)$ as illustrated in the slide.
Continuity of f(x) at a point x_0, means that the mapped point is where it "should be" y_0 = f(x_0)=L (*i.e.*, it has not been excised and moved to some arbitrary location.
Note that the idea of continuity as the function generated by drawing a curve with a pencil without lifting it off the paper is intuitive but imprecise. This is because microscopic irregularities of the paper's surface will lose contact with the pencil leaving "excised points" whose values we are free to define in an arbitrary manner. Thus we need to consider infinitesimal neighborhoods containing an infinite number of points in order to define an accumulation point. The mapping function "pencil(x)" is not continuous in a strict mathematical sense. The concept of a continuous function is an abstraction that is not physically realizable, as is also the case for the related concept of derivative.

1.4.2 Sequences & Limits

Sequences & Limits

Limit of an infinite sequence

Defn: $\{x_k\}_{k=1}^{\infty} = \{x_1, x_2, \cdots x_N, \cdots\}$

If $\exists$ number N such that

$|x_n - x_0| < \varepsilon$ *for all* $n \geq N$ *& all* $\varepsilon > 0$

Then $\lim_{n \to \infty} x_n = x_0$ *and sequence converges to* x_0

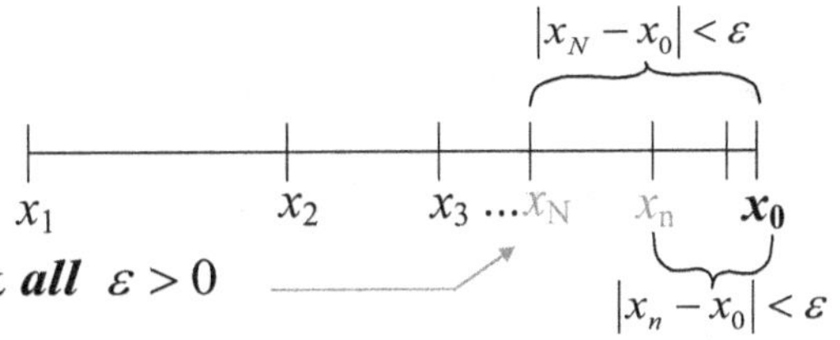

$|x_N - x_0| < \varepsilon$

$x_1 \qquad x_2 \qquad x_3 \dots x_N \quad x_n \quad x_0$

$|x_n - x_0| < \varepsilon$

Continuity for function defined on sequence f(x$_n$)

Thm: *if $\{x_k\}_{k=1}^{\infty}$ is any convergent sequence* $\lim_{n \to \infty} x_n = x_0$

then $\lim_{n \to \infty} f(x_n) = f(x_0) = L$ *(accumulation point)*

i.e., $f(x)$ is continuous at x_0

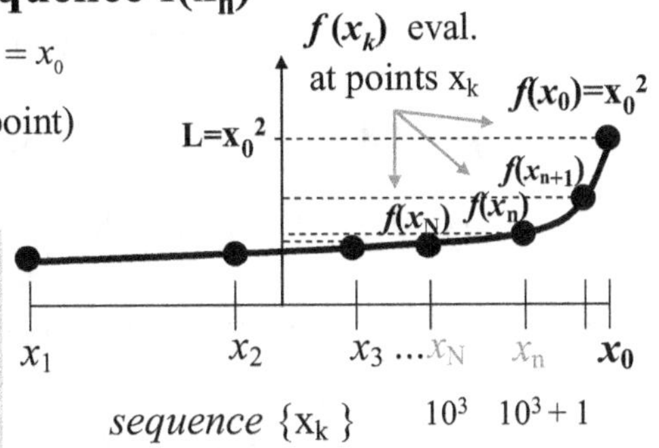

$f(x_k)$ eval. at points x_k $f(x_0) = x_0^2$

$L = x_0^2$

$f(x_{n+1})$

$f(x_N)$ $f(x_n)$

Conv. Seq. Example: $x_n = x_0 - 1/n$; $n = 1, 2, \cdots$

Let $\varepsilon = 10^{-3}$, then $n = N$ is determined by

$|x_N - x_0| = |(x_0 - 1/N) - x_0| = 1/N = 10^{-3}$

Thus for **all** $n > N = 10^3$, [e.g., for $n = N + 1 = 10^3 + 1 > N$]

$|x_n - x_0| = |[x_0 - 1/(10^3 + 1)] - x_0| = 1/(10^3 + 1) < 10^{-3} = \varepsilon$

$\therefore f(x_n) = (x_n)^2 = (x_0 - 1/n)^2 = x_0^2 - 2x_0/n + 1/n^2 \xrightarrow[n \to \infty]{} x_0^2$

$x_1 \qquad x_2 \qquad x_3 \dots x_N \quad x_n \quad x_0$

sequence $\{x_k\}$ 10^3 $10^3 + 1$

Limit of Sequence: Many algorithms involve iterative sequences which converge to a limit solution, so we need to define the limit of an infinite sequence. The illustration shows a (one-sided sequence) which converges to x_0. The definition states that there is a number N=10,000 (say) beyond which the absolute difference between the nth iterate x_n and the limiting value x_0 is less than $\varepsilon < 10^{-3}$ (say). If the latter is true for any ε, then we can say that the limit of the sequence x_n as n approaches infinity is the accumulation point x_0. Note that for simplicity we have illustrated one-sided convergence from the left to right; in general the sequence of iterates will jump to the left and right of the convergent point "x_0".

Continuity for f(x$_n$) defined on a sequence: Given the convergent sequence of iterates $\{x_n\}$, then the mapped points of this sequence $f(\{x_n\})$ form a new sequence $\{f(x_n)\}$ shown in the lower figure. If this new sequence (along the y-axis) also converges to a limit point L in such a way that $\lim_{n \to \infty} f(x_n) = L = f(x_0)$ then limit point is "where it should be" and we say the function $f(x)$ is continuous at the point $x = x_0$. Clearly, if we were to define the mapping function f so that $f(x_0)$ is not equal to L, we produce an excised point in the mapping (at the point x_0) and the function is no longer continuous.

1.4.3 Differentiable Functions & Continuity

Differentiable Functions & Continuity

- Differentiable Function:
  - Limit of secant lines

$$f'(x_0) \equiv \lim_{x \to x_0} \frac{f(x) - f(x_0)}{x - x_0}$$

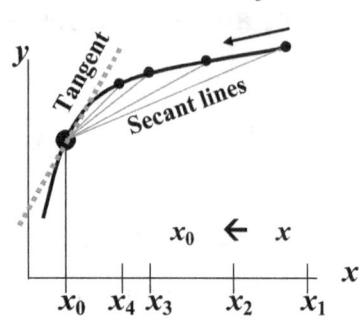

- Thm: Derivative $f'(x_0)$ exists
  $\rightarrow f(x)$ is continuous at x_0
- n^{th} derivative $f^{(n)}(x_0)$ exists
  $\rightarrow f^{(n-1)}(x_0)$ is continuous

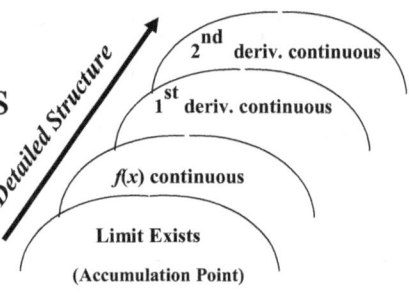

A function is differentiable at x_0 if the slope of the secant line sequence has a limit as $x_n \to x_0$. The existence of the derivative at x_0 implies continuity at x_0 and so on for higher derivatives.

Although existence of a derivative implies continuity the converse is not true; *i.e.*, a function can be continuous at a point and not have a derivative. Consider a "sawtooth" curve with no excised points; it is clearly continuous, yet the derivative at the apex does not exist. This is easy to show since the limit from the left is a positive slope while that from the right is negative slope and hence no unique limit exists at the apex.

The detailed structure of function (mapping) is determined by limit points and derivatives; clearly a random mapping of X➜ Y has no limit points and has no structure. The existence of limit points gives some structure but allows excisions with arbitrary (random) assignments. Requiring the 1st derivative to exist over an (a, b) interval of X yields continuity of the function over that interval and hence gives some structure to the mapping. The existence of each higher order derivative of a function gives a more predictable structure to the functional mapping.

1.4.4 Theorems Relating to Error Estimation

Theorems Relating to Error Estimation

- ## Rollé:

  If $\boxed{f(x) \in C[a,b]}$
  $\boxed{f'(x) \text{ exists } (a,b) \quad (i.e., f(x) \text{ continuous})}$
  $f(a) = f(b) = 0$
  Then $pt \text{ "}c\text{" } \in (a,b) \text{ exists}: \textbf{Tangent} = f'(c) = 0$

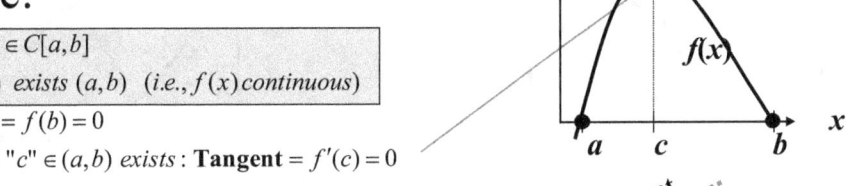

- ## Mean Value (Sec-Tan Thm)

  If $f(x) \in C[a,b]$ *continuous*
  $\quad f'(x) \text{ exists } (a,b)$
  Then $pt \text{ "}c\text{" } \in (a,b) \text{ exists}:$

  $\text{Tangent} = \text{Secant}: f'(c) = \dfrac{f(b) - f(a)}{b - a}$

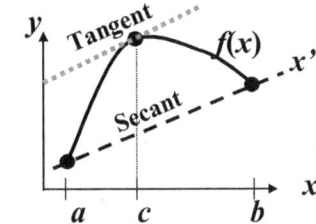

- ## Extreme Value & Corollary:

  If $f(x) \in C[a,b]$ *continuous*
  Then $pts \text{ "}c_1, c_2\text{" } \in [a,b] \text{ exist}: f(c_1) \le f(x) \le f(c_2)$
  $\quad$ If $f'(x) \text{ exists } (a,b)$
  $\quad\quad$ Then $pts \text{ "}c_1, c_2\text{" either"}$
  $\quad\quad$ (i) at end points "a" or "b"
  $\quad\quad$ (ii) where $f'(x) = 0$

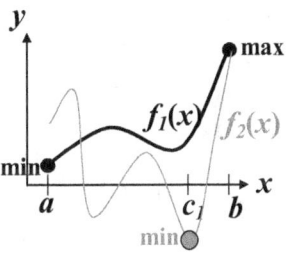

Continuous function theorems:

Rolle: Simply states that if a continuous function f(x) defined on a closed interval [a,b] vanishes at its end points f(a)=f(b)=0, then there must be some intermediate point c at which the derivative f'(c) vanishes. Looking at the top figure, this makes sense since the progressive tangents to the function curve start on the left with a positive slope and end on the right with a negative slope and any intuitive concept of continuity requires that the slope must go through zero at some point in between.

Mean Value (Sec-Tan): This is nothing more than a "Rotated Rolle Theorem" as illustrated in the middle figure where the two end points of the closed interval [a,b] define a new x′–axis and a new closed interval [a*, b*].. Clearly the original function can be re-expressed in the new coordinate system in the form y′ = f*(x′) and it has zeros at its end points f*(a*) = f*(b*) =0; thus Rolle can be applied to this new function to yield the result that its derivative at some point "c*" must be zero f*(x′=c*) =0. Since the geometry cannot depend upon the coordinate system used to view it, we conclude that back in the original coordinate system this geometrically translates to a point c for which the tangent to the curve equals the slope of the x′–axis which is just the secant line between the original points a and b

Extreme Value: This theorem states that for a continuous function there must exist both a minimum (min= f(c₁)) and a maximum (max = f(c₂)) functional value and further that these extremum points c₁ and c₂ occur either at the end points or some where in the interval (a,b) where the derivative is zero. Note: Continuity ensures existence of extrema, since it excludes *excised points* with *arbitrary values*, while existence of the derivative f″(x) in (a,b) "nails down" the location of the extrema to be at the end points or where the derivative is zero.

1.4.5 Weighted Mean Value / Intermediate Value

Wtd. Mean Val. / Intermediate Value

- ## Weighted Mean Value

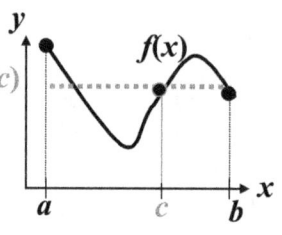

If $f(x) \in C[a,b]$ *continuous*

and $g(x)$ integrable and no sign change

Then *exists* $c \in [a,b]$ such that $f(c) = \dfrac{\int_{x=a}^{b} g(x)f(x)dx}{\int_{x=a}^{b} g(x)dx}$

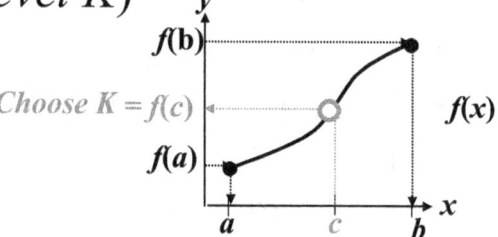

$Average \xrightarrow{g(x)=1} f(c) = \dfrac{\int_{x=a}^{b} f(x)dx}{(b-a)}$

- ## Intermediate Value (*Level K*)

If $f(x) \in C[a,b]$ *continuous*

and choose $"K" : f(a) < K < f(b)$

Then *exists* $c \in (a,b)$

such that $f(c) = K$

"Zero Crossing" $\rightarrow$ Choose K $= 0$

$f(a) < (K=0) < f(b) \Rightarrow f(c) = 0$

Choose $K = f(c)$

f(a) f(b) < 0 is a sufficient condition for a root to exist;
but not a necessary condition

Mean Value: If the weight function is simply the constant g(x)=1, then the theorem states that there is some point "c" for which the area under a box of height f(c) (dashed line in upper figure) times the interval (b-a) must equal the area under the function curve, *i.e.*, Area=(b-a)*f(c) Alternately, the mean value of the function is simply the area divided by the interval f_{mean} = Area/f(c). Extending this concept to arbitrary non-negative g(x), the weighted mean value of the function is just the weighted area divided by the weighted interval.

Intermediate Value: Choose any value K between f(a) and f(b) and there exists a point c ε (a, b) (intermediate point) for which f(c)=K. The connection to root finding is found by setting K=0, since then we are guaranteed a value of c exists whose functional value f(c) = K = 0 (c a root of f(x)) provided K=0 is chosen between f(a) and f(b). The latter statement is written f(a) < 0 < f(b) and thus means that a root exists between a and b provided that the functional values *f(a) and f(b) have different signs.*

Note that the condition *f(a)·f(b)* < 0 is a sufficient condition for a root to exist, but it is not a necessary condition as a root can exist even if the condition is not satisfied. Moreover, there may be more than one root as the function may cross the axis a number of times in between the two points.

Example: Consider the function f(x) = x^2 which is symmetric about the x-axis and has positive values for both a=-1, b=+1 so the condition is not satisfied (no sign difference); however the parabola has a double root at the origin x=0. Now consider its derivative f'(x) = 2 x which is a straight line through the origin; clearly the derivative function f'(x) satisfies the condition: f'(-1)·f'(+1)=-2·(+2) <0.

1.4.6 Generalized Rolle & Taylor

Generalized Rolle & Taylor

- ## Rolle for n=3 (easily generalized)

If $f(x) \in C[a,b]$

$f'(x), f''(x)$ exist on (a,b)

$f(x_1) = f(x_2) = f(x_3) = 0$

Then $f''(c_3) = 0$

Apply Rolle' twice to $f(x)$:

$c_1 \in [x_1, x_2] \, \& \, f(x_1) = f(x_2) = 0 \Rightarrow f'(c_1) = 0$

$c_2 \in [x_2, x_3] \, \& \, f(x_2) = f(x_3) = 0 \Rightarrow f'(c_2) = 0$

Apply Rolle' once to $f'(x)$:

$c_3 \in [c_1, c_2] \, \& \, f'(c_1) = f'(c_2) = 0 \Rightarrow f''(c_3) = 0$

- ## Taylor Polynomial ApproximationTheorem

If $f(x) \in C^n[a,b]$ function & 1st n - deriv cont.

$f^{(n+1)}(x)$ exists (a,b)

Then for $x_0 \in [a,b]$

$$P_n^T(x) \equiv \sum_{k=0}^{n} f^{(k)}(x_0) \frac{(x-x_0)^k}{k!}$$

$$R_n^T = \frac{f^{(n+1)}(\xi)}{(n+1)!} \cdot (x-x_0)^{n+1} \; ; \xi \in (x_0, x)$$

Generalized Rolle: $f(x_1)=f(x_2)=f(x_3)=0$ => Apply Rolle: to pairs of points $\{x_1, x_2\}$ and $\{x_2, x_3\}$; for the first pair Rolle says that there is a "c_1" between x_1 and x_2 for which $f'(c_1)=0$; for the second pair there is a "c_2" between x_2 and x_3 for which $f'(c_2)=0$. Thus the 1st derivative $f'(x)$ is a function that has two zeros at points c_1 and c_2; applying Rolle once again to 1st derivative function we conclude that the 2nd derivative function $f''(x)$ must be zero at new point c_3 between c_1 and c_2, *i.e.,* $f''(c_3)=0$. The generalization to higher derivatives is straightforward. :

Taylor Approximation: Taylor polynomial of degree n approximates the $f(x)$ to terms of degree x^n and the *exact remainder* is given by a formula that involves an unknown point ξ in the interval $[x_0, x]$. The Taylor polynomial and the remainder sum up **exactly** to original function. This is the key expansion used to approximate an analytic function and estimate its truncation error and, for that reason, it is used extensively in formulating numerical algorithms.

1.5 Error Analysis

Error Analysis

- Analysis of Truncation & Round off Errors
- Step Size Dependence of Total Error
- Mathematical and Numerical Instabilities
- Difference Equations
- Reducing Round-Off Errors

The errors generated by a given numerical computational algorithm can be analyzed by employing the simple calculus theorems we have just discussed. Error analysis is a very important aspect of a numerical algorithm since it allows us to separate the total error into two parts, namely, (i) the truncation error (mathematical approximation) and (ii) the round off error (computer operations with finite word size). We shall see that there is a trade-off between the two components since truncation error *decreases* and round off error *increases* as we *decrease* the algorithm stepsize h. We shall see that even if the analytic solution to a problem is stable to parameter perturbations, the numerical algorithm itself can introduce its own unique instabilities. This is because the iterates generated by a numerical algorithm are determined by a recursion relation between new and previous iterates which is equivalent to a difference equation. Therefore, each numerical method has a unique difference equation associated with it and the solutions to these difference equations using fixed digit arithmetic can generate unwanted parasitic terms that diverge from the true solution because of a small initial round off error that multiplies one of the diverging parasitic solutions. Thus error growth is always present and we must always be on the watch for diverging parasitic solutions which can make a seemingly well constructed algorithm diverge exponentially and become useless

1.5.1 Truncation & Round-off Error

Truncation & Round-off Error

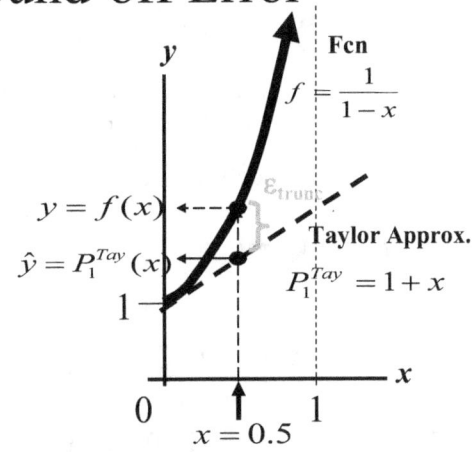

- **Truncation Error** –

    Fcn approx.(Taylor, etc.)

    $$f(x) = P_n^{Tay}(x) + R_n(x)$$

    $$\varepsilon_{trunc} = f(x) - P_n^{Tay}(x)$$

- **Round off Error**

- **(computer word size)**

    - *Inherent:* Input/Output conversion to/from machine representation (quantization)

    - *Propagated:* ε_{RO} computer ops $+ * /$

        - RO after **each** operation to "machine number"

Truncation and Round-off errors are two distinct contributions to errors in numerical algorithms and there is a constant need to make trade-offs so as to minimize their sum.

Truncation error ε_{trunc} is the difference between a function f(x) and its approximating Taylor polynomial $P_{(n)}^{Tay}(x)$. This is illustrated in the slide for the function f(x) =1/(1-x) whose 1st degree (n=1) Taylor polynomial approximation about $x_o = 0$ is given by $P_{(1)}^{Tay}(x)$ =1+x (the dashed straight line).

Round off error is a result of the finite computer word size, *i.e.,* the fixed number of digits used in performing all computer operations. The finite computer word size produces a round-off error each time an input value "x" is converted to a finite machine number representation. After the initial input error, each arithmetic operation (+, /, *) in an algorithm propagates the round-off error because the machine registers can only hold the fixed number of digits in the mantissa; any additional digits are dropped *prior to* performing the next arithmetic operation. The accumulation of round-off errors over many operations can lead to very large errors that would make the final result useless.

1.5.2 Derivation of Truncation & RO Errors

Derivation of Truncation & RO Errors

- 1st Derivative ("Truncate" Taylor Series)

$$f(x_0 + h) = f(x_0) + f'(x_0)h + f''(\xi(x))\frac{h^2}{2}$$

$$x = x_0 + h$$
$$\xi \in [x_0, x_0 + h]$$

- RO Error for each term (Machine precision)

$$f(x_0) = \hat{f}(x_0) + \varepsilon_{sd}(x_0)$$

$$f(x_0 + h) = \hat{f}(x_0 + h) + \varepsilon_{sd}(x_0 + h)$$

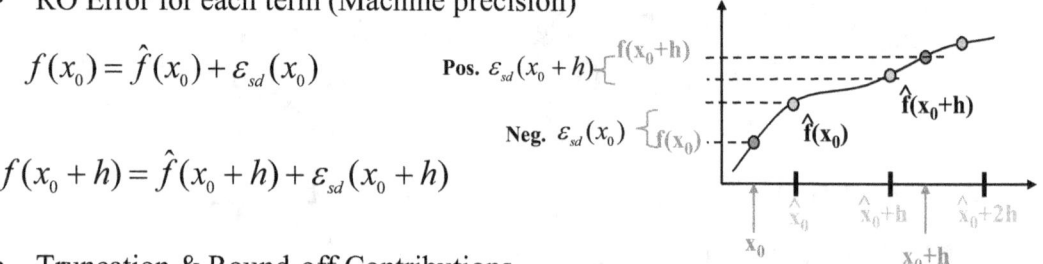

- Truncation & Round-off Contributions

$$f'(x_0) = \underbrace{\frac{\hat{f}(x_0 + h) - \hat{f}(x_0)}{h}}_{\text{Estimated} = \hat{f}'[x_0, h]} + \underbrace{\frac{\varepsilon_{sd}(x_0 + h) - \varepsilon_{sd}(x_0)}{h}}_{\text{RO Error}} - \underbrace{f''(\xi(x))\frac{h}{2}}_{\text{Truncation Error}}$$

- Bounds: $M = \max(f'')$ $e_{sd} = \max(|\varepsilon_{sd}|)$

If we truncate the Taylor series after the linear term in h, then the truncation error is of order h^2 and hence increases quadratically with h. Contributions to the RO error are related to the number of significant digits (sd) and are therefore labeled with an "sd" subscript ε_{sd}. As illustrated on the slide the RO error contribution is the difference between the function evaluated at the desired point $f(x_0)$ and its floating point representation $fl(f(x_0))$ which evaluates the function at a machine point closest to x_0, viz.,

$$\varepsilon_{sd}(x_0) = f(x_0) - fl(f(x_0))$$

The RO error has a different value at $x_0 + h$ than it does at x_0

$$\varepsilon_{sd}(x_0 + h) = f(x_0 + h) - fl(f(x_0 + h))$$

Substituting these expressions into the Taylor formula yields the two distinct contributions to the total error, namely, (i) RO error varying inversely with stepsize h and (ii) truncation error varying as the Taylor truncation term h^2 divided by h as shown in the boxed equation.

Note that the truncation error only depends upon the second derivative f'' at some point ξ in the interval of interest $[x_0, x_0 + h]$ and we find an upper bound to it by taking M= max(f'') in that interval. On the other hand, the RO error subtracts the two round off errors; since the one subtracted might actually be negative making the two add, an upper bound is obtained by adding the absolute values of the two terms; furthermore they generally have different values so we take twice the largest one which we denote e_{sd}. The same logic goes for combining the truncation and RO errors, i.e., we will need to add absolute values of all terms in the boxed equation.

Clearly the opposite behavior of these two error terms with the stepsize h sets up a trade off between decreasing truncation error and increasing RO error as we decrease stepsize h which we explore on the next slide.

1.5.3 Step Size Dependence of Total Error

Step Size Dependence of Total Error

- Total Error $\varepsilon_{tot}(h)$

$$\varepsilon_{tot} \leq \left| f'(x_0) - \hat{f}'[x_0, h] \right| \leq \frac{2e_{sd}}{h} + \frac{Mh}{2}$$

- Optimal Stepsize h_{opt}

$$M = \max(f'')$$
$$= e^x\big|_{x=1} = 2.718$$

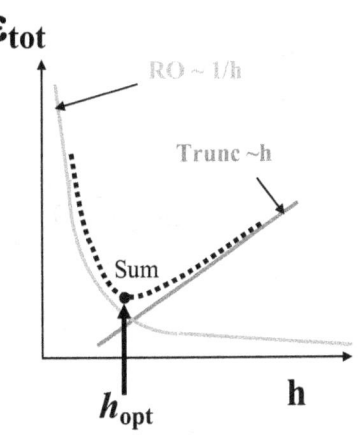

$$h_{opt} = \sqrt{\frac{4e_{sd}}{M}} = \sqrt{\frac{4(5 \cdot 10^{-5})}{2.718}} = .00858$$

$$k = -\frac{\log_{10}(.00858)}{\log_{10}(2)} = 6.86 \rightarrow \boxed{7}$$

The RO errors may have different signs, but the worst case occurs when the two RO errors add. Thus, we must take the sum of their absolute values and then maximize that sum by taking twice the largest contribution. The absolute value of the second derivative term is added to the latter sum to obtain the worst case total error in the upper boxed equation.

The total error is therefore the sum of two positive terms, namely, (i) the RO error term (green) which varies inversely with h and (ii) the truncation error (red) which increases linearly with h. The optimum value h_{opt} shown in the figure is computed by setting the first derivative of the total error $\varepsilon_{tot}(h)$ equal to zero to yield the value shown. If we now compare this result with the MatLab simulation used to estimate the values of numerical 1st derivative of e^x as a function of decreasing step size (Slide#1-6), we see that it is precisely the trade-off between truncation and round off error that gives the optimum value for the halving index $k \approx 7$ which is close to k=8 found in Slide#1-6 by direct calculation.

Furthermore, the drop to zero results when the stepsize h is zero to 5 significant digits; this occurs at k=15 where $h_5 = 3.05 \times 10^{-5} < 5 \times 10^{-5}$ (see Slide#1-14 machine epsilon). Thus, because the two functional evaluations $f(x_0 + h_5)$ and $f(x_0)$ are equal to within 5 sd, subtractive cancellation causes all significant digits to be lost resulting in a value of zero for the derivative.

1.6 *Mathematical (Inherent) Instability*

Mathematical (Inherent) Instability

Sensitivity Near a Singularity

$$y = f(x) = \frac{1}{1-x}$$

Inverse x=x(y)
y(1-x)=y →
x=(y-1)/y

Taylor
Poly.
approx.

$$\hat{f} = 1 + x$$

$$y = f(x = .5) = 2$$

Δy

$$\hat{y} = \hat{f}(x = .5) = 1.5$$

$$\hat{x} = f^{-1}(\hat{y}) = (\hat{y} - 1)/\hat{y}$$
$$= (1.5 - 1)/1.5 = .333$$

1

0

1

$$\hat{x} = f^{-1}(\hat{y}) = .333$$

Δx

$$x = 0.5$$

Condition Number

K(x) determines how much a small relative change in input $|\Delta x/x|$ is magnified in the output $|\Delta y/y|$, viz.,

$$\left|\frac{\Delta y}{y}\right| = K(x)\left|\frac{\Delta x}{x}\right|$$

$$K(x) = \left|\frac{\Delta y / y}{\Delta x / x}\right| = \left|\frac{f'(x)}{f(x)} \cdot x\right|$$

$$= \left|\frac{-1/(1-x)^2}{1/(1-x)} \cdot x\right| = \left|\frac{x}{(1-x)}\right|$$

It has been pointed out that we must avoid physical models with singular points such as x=1 for the function 1/(1-x). Although the singularity is obvious by inspection of its graph for this simple function, a more complicated function requires a non-graphical characterization of the mathematical instability and leads to the definition of the "condition number" K(x) given in the boxed equation.

The condition number, K(x) expresses how much a *small relative change* in the input [$\Delta x/x$] is *magnified in the output* [$\Delta y/y$] , viz.,

$$[\Delta y/y] = K(x) * [\Delta x/x].$$

Δy is the difference y − y_{cap} between the exact function and its 1st degree Taylor polynomial "1+x"; similarly Δx is the difference x − x_{cap}, where x_{cap} is the pull-back using the exact function f $^{-1}(y_{cap})$ (see figure)

$$f^{-1}(y_{cap}) = (y_{cap}-1) / y_{cap}$$

The grey box on the right solves for the condition number K(x) and upon setting $\Delta y = f'(x)\Delta x$, and substituting both the function f(x) and its derivative f'(x) = -1/(1-x²) yields the condition number for our simple example

$$K(x) = |x/(1-x)|$$

The above expression becomes large as x ➔1 and is thus clearly ill-conditioned near x=1. The boxed definition for K(x) is suitable for assessing the condition number without the need for plotting graphs.

1.6.1 Sensitivity & Condition Number-Details

Sensitivity & Condition Number-Details

$$y = f(x) = \frac{1}{1-x}$$

$$\hat{y} = f(\hat{x}) \cong f(x) + f'(x)(x - \hat{x})$$

1st Order Taylor Expansion

$$|\Delta y| = |f'(x)| \cdot |\Delta x|$$

$$K(x) = \left| \frac{\Delta y / y}{\Delta x / x} \right| = \left| \frac{\Delta y / \Delta x}{y} \cdot x \right| = \left| \frac{f'(x)}{f(x)} \cdot x \right|$$

Condition Number measures %change of output to input

$$= \left| \frac{-1/(1-x)^2}{1/(1-x)} \cdot x \right| = \left| \frac{x}{(1-x)} \right|$$

Highly sensitive (ill-conditioned) near x=1

Here are the details of the Taylor expansion and the identification of terms in the computation of condition number on the previous slide. A 1st order Taylor expansion about x_{hat} is given in the second line and the third line results from the definitions $\Delta y = y_{hat} - f(x_{hat})$ and $\Delta x = x - x_{hat}$. Finally writing down the definition of the condition number $K(x)$ and substituting the expressions for the $y = f(x)=1/(1-x)$ and $\Delta y = f'(x) \Delta x = -1/(1-x)^2$ we obtain the desired result.

1.7 *Numerical Instability*

Numerical Instability

- Even if the mathematical function is not sensitive to small input changes, the numerical algorithm may have instabilities of its own

- Numerical solutions introduce "parasitic" solutions to the difference equations that bear no relation to the true solution and may cause algorithm to become unstable.

- Error growth:

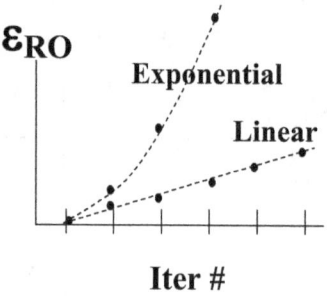

The fascinating thing about numerical methods is that, even when there are *no inherent singularities* in the mathematical formulation of the problem, iterative numerical algorithms may in fact generate *parasitic solutions* that diverge exponentially with the iteration index k.

This is because a given numerical algorithm can be re-expressed in a number of ways as a difference equation involving various combinations of it's iterates y_{k-2}, y_{k-1}, y_k, y_{k+1} etc. The solution of a 2nd or higher order difference equation generates multiple solutions; it generates the correct solution together with a number of parasitic solutions that are unrelated to the original problem.

The general solution to a difference equation is a linear combination of the correct solution and all of the parasitic solutions. In terms of the original problem being solved, the *parasitic components represent error terms* that must be controlled; for if just one of these parasitic solutions diverges, it will dominate and thus mask the correct solution to the problem.

Since there is no unique difference equation to a given problem, we must choose our algorithm carefully in order to control the error growth of the parasitic components. The next two slides illustrate this point.

1.7.1 Solution to a Difference Equation

Solution to a Difference Equation

- **2nd DfE:** $\quad p_k = (10/3)p_{k-1} - p_{k-2}$

- **Two IC:** $\quad k=0: p_0 = 1 \; ; \; k=1: p_1 = 1/3$

- **Trial Soln:** $\quad p_k = \lambda^k$

- **Roots to Quadratic Eqn** $\quad (\lambda^2 - (10/3)\lambda^1 + 1)\lambda^{k-2} = 0$

$$\lambda_1 = 1/3 \; ; \; \lambda_2 = 3$$

- **Lin. Comb. = General Soln:** $\quad p_k = C_1(\lambda_1)^k + C_2(\lambda_2)^k$

$$p_k = C_1 * (1/3)^k + C_2 * (3)^k$$

- **Apply IC to Obtain Soln:**

$$p_0 = 1 = C_1 * (1/3)^0 + C_2 * (3)^0 \qquad \Rightarrow C_1 = 1$$
$$p_1 = 1/3 = C_1 * (1/3)^1 + C_2 * (3)^1 \qquad C_2 = 0$$

$$p_k = (1/3)^k$$

Shown is a typical 2nd order difference equation with two initial conditions. The solution method is analogous to that used for a 2nd order differential equation; we substitute a trial solution $p_k = \lambda^k$ (for ODE we would make a similar substitution $y = e^{mt}$) into the difference equation. This yields a quadratic in λ with two roots $\lambda_1 = 1/3$ and $\lambda_2 = 3$.

The general solution is written as a linear combination of the trial solutions associated with these two root solutions as shown. Applying the initial conditions (***using exact arithmetic***) yields the two constants $C_1 = 1$ & $C_2 = 0$ and hence the solution to the original problem
$$p_k = 1*(1/3)^k + 0*(3)^k = (1/3)^k.$$
This solution converges to zero as it should since the term 3^k has coefficient exactly equal to zero. However, the next slide shows how limited numerical precision of finite digit arithmetic leads to a solution in which both roots contribute and thus leads to numerical instability .

1.7.2 Example of Numerical Instability

Example of Numerical Instability

Difference Equation	Solution
1) $p_k = (1/3) * p_{k-1}$; $k = 1, 2, \cdots$ $IC: p_0 = 1$	$p_k = (1/3)^k$; $k = 1, 2, \cdots$
2) $p_k = (10/3)p_{k-1} - p_{k-2}$; $k = 2, 3, \cdots$ $IC: \begin{Bmatrix} p_0 = 1 \\ p_1 = 1/3 \end{Bmatrix} \Rightarrow \begin{Bmatrix} C_1 = 1 \\ C_2 = 0 \end{Bmatrix}$ $\begin{Bmatrix} p_0 = 1.0000 \\ p_1 = .33333 \end{Bmatrix} \Rightarrow \begin{Bmatrix} C_1 = 1.0000 \\ C_2 = -.12500 \times 10^{-5} \end{Bmatrix}$	$p_k = C_1 * (1/3)^k + C_2 * (3)^k$ $p_k = 1 \cdot (1/3)^k + 0 \cdot (3)^k$ $p_k = 1.0000 \cdot (1/3)^k - \underbrace{.12500 \times 10^{-5} \cdot (3)^k}_{\text{exponential growth}}$

Computational Details: Imposing IC to 5 significant digits, we have explicitly

$$(1)\ \hat{p}_0 = \boxed{1.0000} = C_1(\tfrac{1}{3})^0 + C_2(3)^0 = \boxed{C_1(1.000\overline{0}...) + C_2(1.000\overline{0}...)}$$

$$(2)\ \hat{p}_1 = \boxed{0.33333} = C_1(\tfrac{1}{3})^1 + C_2(3)^1 = \boxed{C_1(0.33333...) + C_2(3.000\overline{0}...)}$$

Forming the linear comb. $\{(0.3333\overline{3}...) \times (1) - (2)\}$ yields : | Similarly, $\{(3.000\overline{0}...) \times (1) - (2)\}$ yields:

$0.(3333\overline{3}...) \times (1.0000) - .33333 = (0.3333\overline{3}... - 3.000\overline{0}...) \times C_2$

$$C_2 = \frac{.33333 \times 10^{-5}}{-2.6667} = -.12500 \times 10^{-5} \qquad\qquad C_1 = \frac{3.000\overline{0}... - .33333}{2.6667} = .10000 \times 10^1$$

The table shows 1st and 2nd order difference equations that can be used to solve the same problem.
1st order difference equation has the solution shown with no parasitic terms and is stable. It is solved by simply writing down the sequence of equations
$$\{p_1 = 1/3\ p_0,\ p_2 = 1/3\ p_1,\ p_3 = 1/3\ p_2, \ldots\ p_{n-1} = 1/3\ p_{n-2},\ p_n = 1/3\ p_{n-1}\}$$
taking their product, canceling common terms from the two sides, and applying the IC $p_0 = 1$, *viz.*,
$$(p_1\ p_2\ p_3 \ldots\ p_{n-2}\ p_{n-1})\ p_n = (1/3)^n\ p_0 (\ p_1\ p_2\ p_3 \ldots\ p_{n-2}\ p_{n-1}) \rightarrow p_n = (1/3)^n\ p_0 = (1/3)^n$$
Alternately, substitution of $p_k = \lambda^k$ into the difference equation yields the single root $\lambda = 3$ and hence the general solution $p_n = C\ (1/3)^n$ which, upon applying the IC $p_0 = 1$ again yields $p_n = (1/3)^n$.
2nd order difference equation was solved on the previous slide using *exact arithmetic* and yielded solution #1. However, numerical algorithms operate with a limited precision determined by the computer word size. In the bottom panel of the slide we show the explicit calculation for the coefficients C_1 and C_2 for the case of 5 significant digits. We see that a parasitic term eventually dominates the solution even though it has a very small coefficient of $-.12500\ 10^{-5}$. For large k the term $-.12500\ 10^{-5} *3^k$ diverges "exponentially" and therefore leads to an unstable solution. Parasitic terms are always present in the solution to 2nd and higher order difference equations and must be tested for roots that lead to divergent solutions.

1.8 *Reducing Round-Off Error*

Reducing Round-Off Error

Reduce Number of Operations & Avoid Cancellations

1. New Method of computation
   a. Rationalize (add rather than subtract)
   b. Trigonometric or geometric identities
   c. Expand in Taylor series
2. Double Precision Subtraction
3. Sequence of Operations
   a. Nesting polynomials –Horner's method
   b. Add small numbers first; otherwise "lost" in large numbers

As seen in the last slide small round off errors cannot be ignored if the numerical technique involves a recursive iteration procedure because of the possible existence of divergent parasitic terms in the solution to the difference equation.

Round off errors are a fact of life in *all numerical algorithms* (iterative or not) because the round off error that occurs after each and every operation (multiply, add, divide) in finite digit arithmetic will result in a loss of significant digits (sd). Moreover, as the number of operations becomes large we have seen (see derivative example on Slides#1-5, 1-6) that the algorithm can become completely useless even if well-formulated mathematically.

We list a few representative techniques to reduce round-off errors, namely, (i) introducing a new method of computation that reduces the number of operations, (ii) invoking double precision to avoid subtractive cancellation, or (iii) changing the sequence of operations. We shall look at a few examples in detail.

1.8.1 New Method of Computation -1

New Method of Computation -1

- ## *Quadratic Equation*

$$x^2 + 62.10x + 1 = 0 \qquad \left(x_+\right)_{exact} = -0.01610723 \, ; \qquad \hat{x}_+ = -0.02000$$

$$x_\pm = \frac{-b \pm \sqrt{b^2 - 4ac}}{2a} \qquad \left(x_-\right)_{exact} = -62.08390 \, ; \qquad \hat{x}_- = -62.080 \quad \checkmark$$

- ## *Rationalize Expression for x_+*

$$x_+ = \frac{-b + \sqrt{b^2 - 4ac}}{2a} \cdot \frac{-b - \sqrt{b^2 - 4ac}}{-b - \sqrt{b^2 - 4ac}}$$

$$= \frac{-2c}{b + \sqrt{b^2 - 4ac}} = \frac{-2.0000}{62.10 + 62.06} = -.01610 \quad \checkmark$$

In a direct application of the quadratic formula we find that 5 sd arithmetic yields the larger root with 4 sd accuracy, but yields only 1 sd accuracy for the smaller root. The problem is that b is so large that its square completely dominates the expression inside the radical, so that we wind up subtracting nearly equal numbers b −√(b²-4ac) ≈ b − √(b²) ≈ 0 and thus lose almost all sd. The cure is to rationalize the expression by multiplying it by a convenient form of "1" which replaces the difference in the numerator by a sum in the denominator; computation using this rationalized expression yields the smaller root to 4 sd as shown in the slide.

1.8.2 New Method of Computation -2

New Method of Computation -2

- **Polynomial** $P(x) = x^3 - 3x^2 + 3x - 1$

- **3^{sd}** $(2.19)^2 = 4.7961 \Rightarrow \underline{4.80}$; $(2.19)^3 = 2.19*(2.19)^2 = 2.19*4.80 = 10.512 \Rightarrow \underline{10.5}$

$$P(2.19) = 10.5 - \underbrace{3*4.80}_{14.4} + \underbrace{3*2.19}_{6.57} - 1 = \boxed{1.67} \longleftarrow \Delta=.015$$

$$\underbrace{}_{-3.90} \qquad \underbrace{}_{5.57}$$

$$P_{exact}(2.19)=1.6852$$

$\Delta=.005$

- **Nested Polynomial** $P(x) = ((x-3)x+3)x - 1$

$$P(2.19) = ((2.19-3)*2.19+3)*2.19-1 = \boxed{1.69}$$

$$\underbrace{}_{-.810}$$
$$\underbrace{}_{-1.77\ 39}$$
$$\underbrace{}_{1.23}$$
$$\underbrace{}_{2.69\ 37}$$

Direct polynomial evaluation leads to subtractive cancellation and loss of 1 sd (significant digit). Nesting the polynomial reclaims the 3rd sd.! Note that in the 3 sd computations on this slide we carefully round off (chop to exactly 3 digits) *after each and every arithmetic operation* just as a 3 digit computer must do. The numbers under the braces of each binary operation show the full result which is then chopped to 3 sd before going on to the next operation. Although the effect in this example is quite small the round off error for large degree polynomials can be devastating!

1.8.3 New Method of Computation -3

New Method of Computation -3

- **Exact Fcn**

$$f(x) = \frac{e^x - 1 - x}{x^2}$$

$$f(.01) = \frac{e^{.01} - 1 - .01}{.01^2} = \frac{1.01005 - 1.01}{.0001} = .500000$$

- **Polynomial Approx.**

Cancel Mathematically

1^{sd}

$$f(x) \cong P_{(4)}(x) = \frac{\{1 + x + x^2/2 + x^3/6 + x^4/24\} - 1 - x}{x^2}$$

$$= \frac{1}{2} + \frac{x}{6} + \frac{x^2}{24}$$

$f_{exact}(.01) = .5016708$

$$P_{(4)}(.01) = .5 + \frac{.01}{6} + \frac{.01^2}{24} = .501671$$

6^{sd}

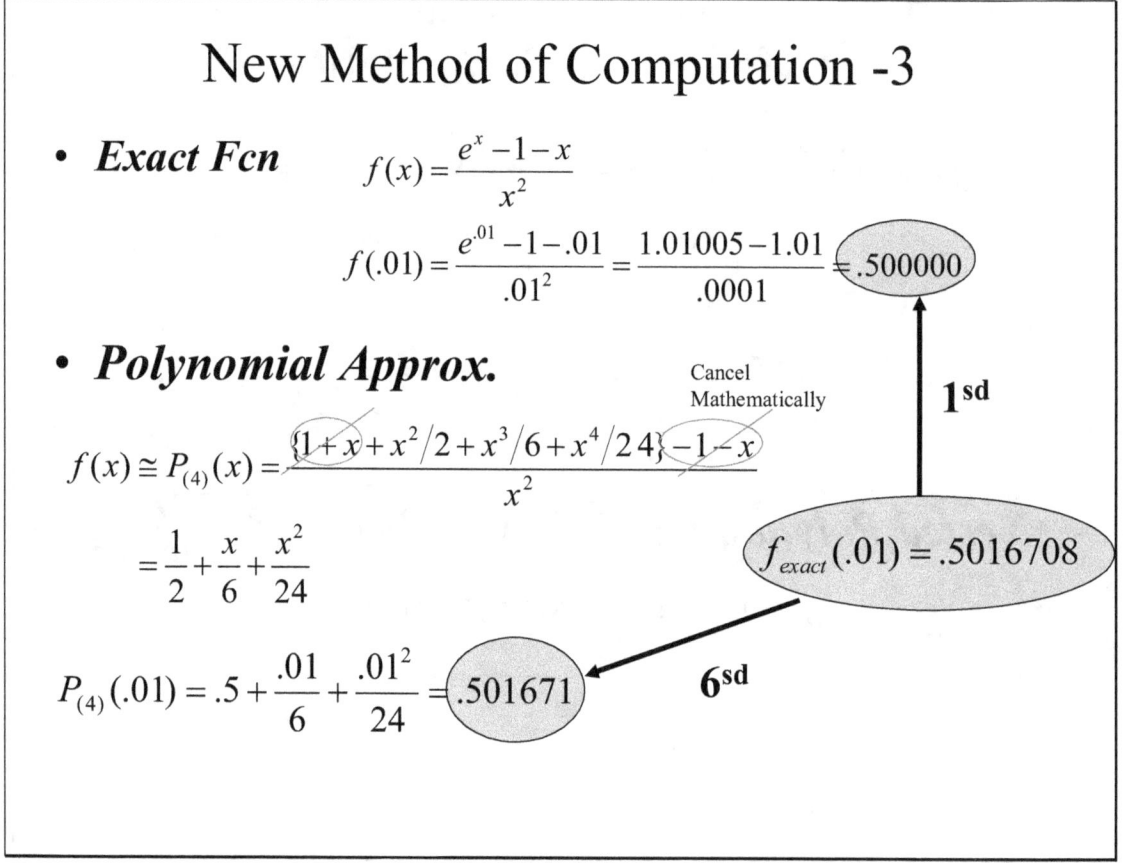

Evaluating the exponential term to 6 sd and subtracting (1+x) yields a result with only 1 sd. However, using a Taylor polynomial for the exponential instead, allows the two large terms (1+x) and (-1-x) to cancel leaving a sum of small numbers to be added without cancellation. The result is good to 6 sd !!

1.8.4 Double Precision Subtraction

Double Precision Subtraction

- ## 5 digit chopping loses all 5 significant digits

$$(\tfrac{1}{3} - .33332) \times 53678 = \underbrace{.71570}\ 6648774$$

$$\underbrace{(.33333\cdots - .33332)}_{.00001} \times 53678 = \underbrace{.53678}$$

no significant digits

Exact $= .715707$

- ## Double precision subtraction

$$\underbrace{(.33333\ 33333\cdots - .33332\ 00000)}_{=.13333\,00000\times10^{-4}} \times 53678$$

$$(.13333\,00000\times10^{-4}) \times .53678 \times 10^{5} = \underbrace{.71569}$$

4-5 significant digits

Here again subtractive cancellation of the two terms in the parentheses losses 4 sd; the situation is compounded by the large multiplier "53678" to give a completely erroneous answer.

Judicious use of *double precision* to do the subtraction and then changing back to single precision for the multiplication yields a result good to 4-5 sd.

2 Root Finding Techniques

2 - Root Finding Techniques

- Bisection, False Position
- Secant, Newton-Raphson
- Fixed Point Theorem
- Hybrid Techniques

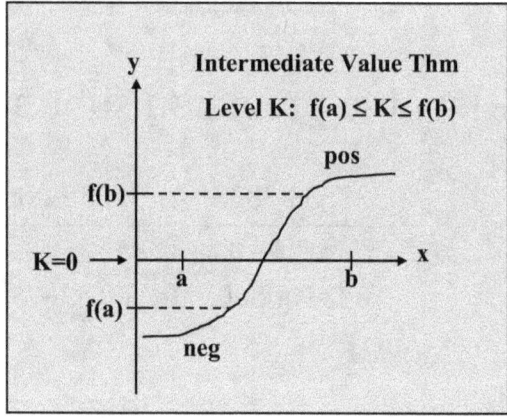

All root finding techniques rely on the Intermediate Value Theorem (IVT) to determine the zero-level crossings (K=0) of a function f(x). The theorem states that for a continuous function f(x) on [a,b] with a well-defined first derivative in the open interval (a,b), there is a point c in that interval for which f(c) = 0, (*i.e.,* a root) provided that the functional values f(a) and f(b) have different signs at the interval endpoints. This theorem states a sufficient condition for the existence of a root; however, a root may also exist even if the theorem conditions are not satisfied (not a necessary condition).

There are many methods for root finding and there are trade-offs between their rates of convergence, their ability to find the root in a specific interval, and their ability to handle multiple roots and complex roots.

2.1 *Iterations & Stopping Criteria*

Iterations & Stopping Criteria

- Generate sequence $\{p_n\}_{n=1}^{\infty}$

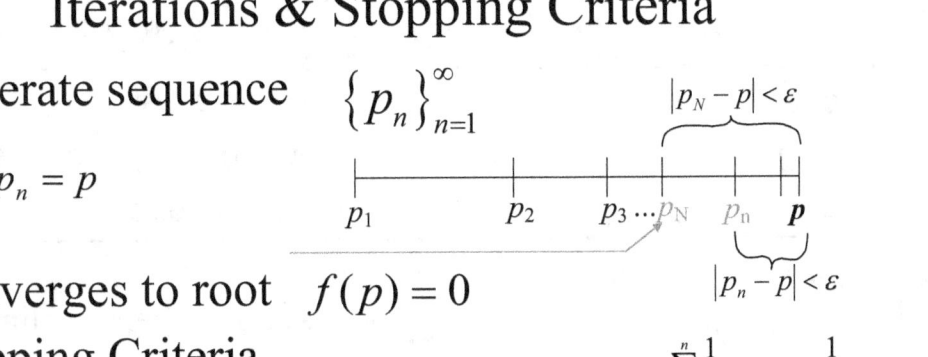

$$\lim_{n\to\infty} p_n = p$$

- Converges to root $f(p) = 0$
- Stopping Criteria
  - Absolute $\boxed{|p_n - p_{n-1}| < \varepsilon}$ Log Divergence

  ✓Relative $\boxed{\dfrac{|p_n - p_{n-1}|}{|p_n|} < \varepsilon}$

  - Functional Value $\boxed{|f(p_n)| < \varepsilon}$

$|p_n - p| < \varepsilon$

$$p_n = \sum_{k=1}^{n} \frac{1}{k} = p_{n-1} + \frac{1}{n}$$

$$|p_n - p_{n-1}| = \left|\frac{1}{n}\right| < \varepsilon$$

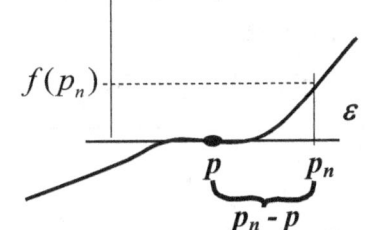

Numerical root finding techniques generate a sequence of estimates $\{p_1, p_2, \ldots, p_k\}$ which converge to the desired root "p" . The algorithm also requires a stopping criterion to determine when we have achieved the desired accuracy tolerance, say $\varepsilon < 10^{-6}$.

There are three typical stopping criteria:

i) **absolute difference** between two consecutive iterates,
ii) **relative error** (ratio of absolute difference over the last iterate),
iii) **absolute value** of the function

Criteria i) can fail *via* logarithmic divergence, even if the absolute differences become vanishingly small

Criteria iii) can fail for very flat functions with many derivatives being zero at the root value

Criteria ii), the relative error, is usually the best choice, but may be used in conjunction with the other two.

2.2 Root Finding Overview

Root Finding Overview

Int.Value Thm: $f(a) \cdot f(b) < 0 \Rightarrow c \in [a,b]; \quad f(c) = 0$

1 *Bracketing Methods*	– Requires a test $\quad f(a) \cdot f(b) < 0$
– Bisection	– Slow Convergence (linear)
– False Position	– Always finds a root
2 *Non-Bracketing*	– No test
– Secant	– Fast Convergence (>linear)
– Newton-Raphson	– Must be near root
– Fixed Point	– May find undesired root
– Mueller	
3 *Hybrid*	– Start with slow Bracketing
– Combine 1 & 2	– Then fast "Turbo"

Methods for finding a root in a given interval [a,b] falls into three categories as follows(See MatLab script on Slides#5-10 to 5-12)

1) Bracketing methods which always find a root in the desired interval, but are slow to converge and require a functional value test to determine the new interval

2) Non-bracketing methods which are fast and require no test, but must start out near the root and even then may not find a root in the desired interval

3) Hybrid is a combination of 1) and 2) which starts with bracketing and then switches to a bracketing method according to some criterion. This method uses a faster non-bracketing method to increase the accuracy once we are certain that we are closing in on the desired root.

2.3 *Bracketing Methods - Bisection*

Bracketing Methods - Bisection

- **Bisection**
  - Halve $p_1 = \dfrac{a_1 + b_1}{2} = \dfrac{a + b}{2}$
  -
  - Test $f(a_n) \cdot f(b_n) < 0$
  - n^{th} iterate

$$p_n = \frac{a_n + b_n}{2}$$

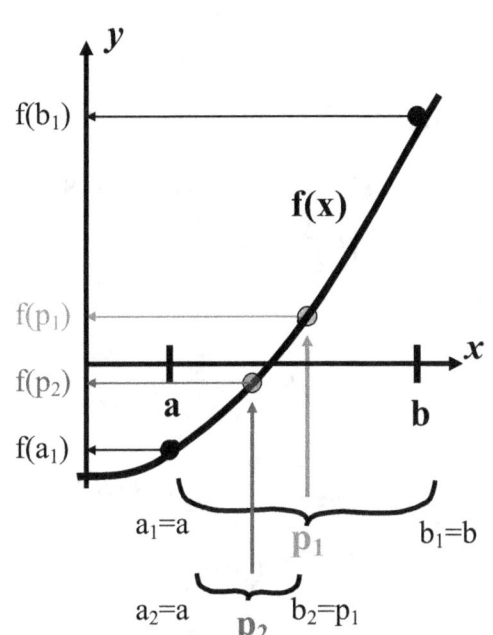

First an interval $[a_1, b_1]$ for which $f(a_1) * f(b_1) < 0$ establishes the existence of a root. Then the interval is divided in half and the 1st root estimate p_1 is simply the average $(a_1 + b_1)/2$. Each half is tested for sign change of the endpoint functional values, thus establishing a new smaller interval $[a_2, b_2]$ for the root. The 2nd root estimate is

$$p_2 = (a_2 + b_2)/2;$$

and the iterations continue until the stopping criterion is satisfied.

2.3.1 Bracketing Methods – False Position

Bracketing Methods – False Position

False Position

— Intercept w/x-axis

$$\frac{f(b_1) - f(a_1)}{b_1 - a_1} = slope = \frac{f(b_1) - 0}{b_1 - p_1}$$

$$p_1 = b_1 - \frac{f(b_1) \cdot (b_1 - a_1)}{f(b_1) - f(a_1)}$$

— Test $\quad f(a_n) \cdot f(b_n) < 0$

— n^{th} iterate

$$p_n = b_n - \frac{f(b_n) \cdot (b_n - a_n)}{f(b_n) - f(a_n)}$$

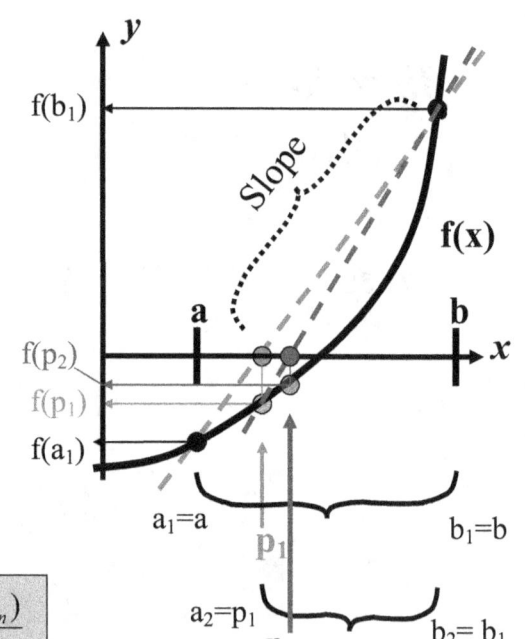

First an interval $[a_1, b_1]$ for which $f(a_1)* f(b_1) < 0$ establishes the existence of a root. Instead of simply bisecting the interval to find a root estimate, we use the intersection of the secant line with the x-axis to determine the next iterate. This secant slope clearly takes into account the curvature of the function and hence is more specific to the function than is a simple bisection the interval.

Then a secant line is drawn from $(a_1, f(a_1))$ to $(b_1, f(b_1))$ and its intersection with the x-axis determines the first iterate

$$p_1 = b_1 - [f(b_1)*(b_1 - a_1)] / [f(b_1) - f(a_1)]$$

which divides the original interval into two unequal intervals. Each interval is tested for sign change of the endpoint functional values, thus establishing a new smaller interval $[a_2, b_2]$ for the root. The 2nd root estimate is

$$p_2 = b_2 - [f(b_2)*(b_2 - a_2)] / [f(b_2) - f(a_2)] ;$$

and the iterations continue until the stopping criterion is satisfied.

2.4 *Non-Bracketing – Secant*

Non-Bracketing – Secant

Secant Method

– Intercept w/x-axis

$$slope = \frac{f(p_1) - f(p_0)}{p_1 - p_0} = \frac{f(p_1) - 0}{p_1 - p_2}$$

– Two Most Recent Pts

(No test!)

$$p_2 = p_1 - \frac{f(p_1) \cdot (p_1 - p_0)}{f(p_1) - f(p_0)}$$

– n^{th} iterate

$$p_n = p_{n-1} - \frac{f(p_{n-1}) \cdot (p_{n-1} - p_{n-2})}{f(p_{n-1}) - f(p_{n-2})}$$

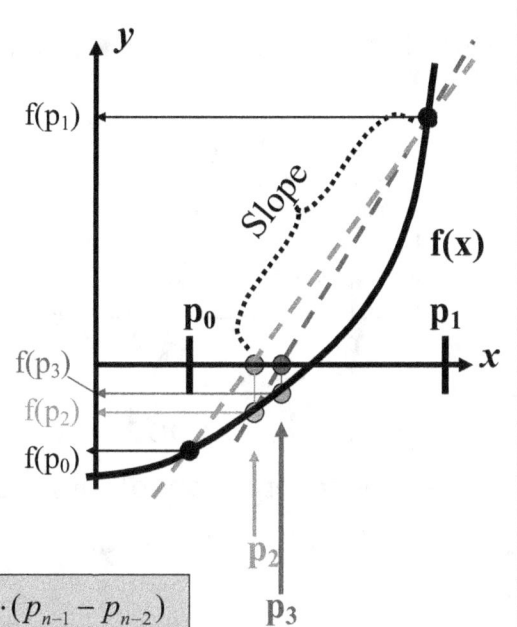

An interval $[a_1, b_1]$ for which $f(a_1) * f(b_1) < 0$ establishes the existence of a root. Or we can simply pick two arbitrary points $\{p_0, p_1\}$ near an expected root without actually testing for a root.

Then a secant line is drawn between the first two points from $(p_0, f(p_0))$ to $(p_1, f(p_1))$ and its intersection with the x-axis determines the 2nd iterate

$$p_2 = p_1 - [f(p_1) * (p_1 - p_0)] / [f(p_1) - f(p_0)],$$

which leaves us with the sequence $\{p_0, p_1, p_2\}$. No test is made and the next iterate uses the "two most recent iterates" $\{p_1, p_2\}$ and applies the same formula with all indices incremented by 1 to yield

$$p_3 = p_2 - [f(p_2) * (p_2 - p_1)] / [f(p_2) - f(p_1)] ;$$

the iterations continue until the stopping criterion is satisfied.

Note that because no tests are made, the root may not exist at all or may be outside the original interval $[p_0, p_1]$; but the convergence is fast and we will find out very quickly whether or not we are finding the desired root.

2.4.1 Extension of Secant – Muller's Method

Extension of Secant – Muller's Method

Most Recent 2 Pts ➜ Most Recent 3 Pts (Quadratic)

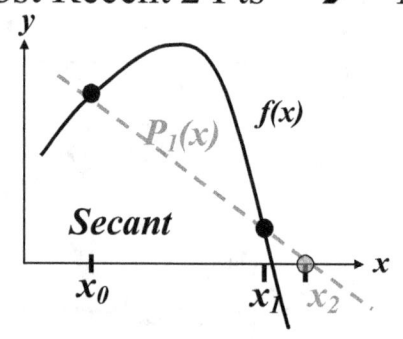

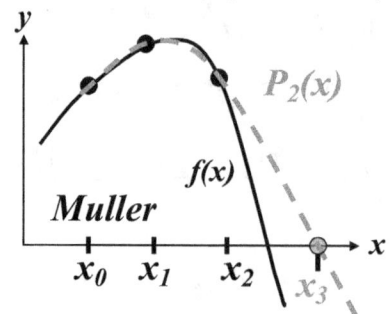

- Muller's Method:
  - Converges rapidly for arbitrary initial values
  - Yields complex roots

Muller's Method is a generalization of the secant method that starts with three points (instead of 2) and fits a quadratic curve (instead of a straight line) through the three initial points. This quadratic curve fitted through these three points intersects the x-axis in at most 2 points and we choose the point x_3 that is closest to x_2 yielding the sequence $\{x_0, x_1, x_2, x_3\}$. We then proceed (without tests) using the three most recent points $\{x_1, x_2, x_3\}$ to generate a new quadratic curve and new intersection x_4. This process continues until the stopping criterion is satisfied. The detailed algorithm is given on the next slide.

2.4.2 Muller's Method Details

Muller's Method Details

Quadratic through 3 pts: $(x_0, f(x_0))$; $(x_1, f(x_1))$; $(x_2, f(x_2))$;

$$P_2(x) = a(x - x_2)^2 + b(x - x_2) + c$$

Fit Coeffs:

$$a = \frac{-(x_0 - x_2)[f(x_1) - f(x_2)] + (x_1 - x_2)[f(x_0) - f(x_2)]}{(x_0 - x_2)(x_1 - x_2)(x_0 - x_1)}$$

$$b = \frac{+(x_0 - x_2)^2[f(x_1) - f(x_2)] - (x_1 - x_2)^2[f(x_0) - f(x_2)]}{(x_0 - x_2)(x_1 - x_2)(x_0 - x_1)}$$

$$c = f(x_2)$$

Find Approx. Root $f(x) \cong P_2(x)$

$$0 = P_2(x_3) = a(x_3 - x_2)^2 + b(x_3 - x_2) + c$$

First Iterate:

$$x_3 = x_2 + \frac{-2c}{b + signum(b)\sqrt{b^2 - 4ac}}$$

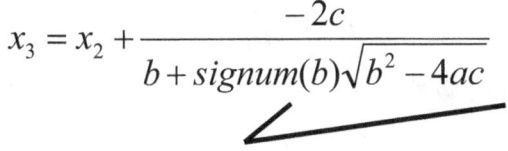

Choose root close to x_2

The quadratic form $P_2(x)$ is evaluated at the three initial points $\{(x_0, f(x_0)), \{(x_1, f(x_1)), \{(x_2, f(x_2))\}$ to give three equations in three unknowns; their solution yields the three formulas for the unknown coefficients a, b, and c given on the slide. These coefficients define the quadratic $P_2(x; a,b,c) = 0$ whose two roots are illustrated in the figure by the intersections of the red dashed curve with the x-axis. We choose the root x_3, closest to the last point x_2 by using the "sign function of b" (signum(b)) to make sure that the two terms in the denominator always add to make it large in absolute value and hence the correction term smallest. The result is given by the formula at the bottom of the slide. The next step is to drop x_0 from our list and use the three most recent points, *i.e.*, $\{x_1, x_2, x_3\}$ to compute a "new" set of quadratic coefficients a, b, and c, and again solve the quadratic $P_2(x; a,b,c) = 0$ to find the new root x_4; this process continues until the convergence criterion is met.

2.4.3 Non-Bracketing – Newton-Raphson

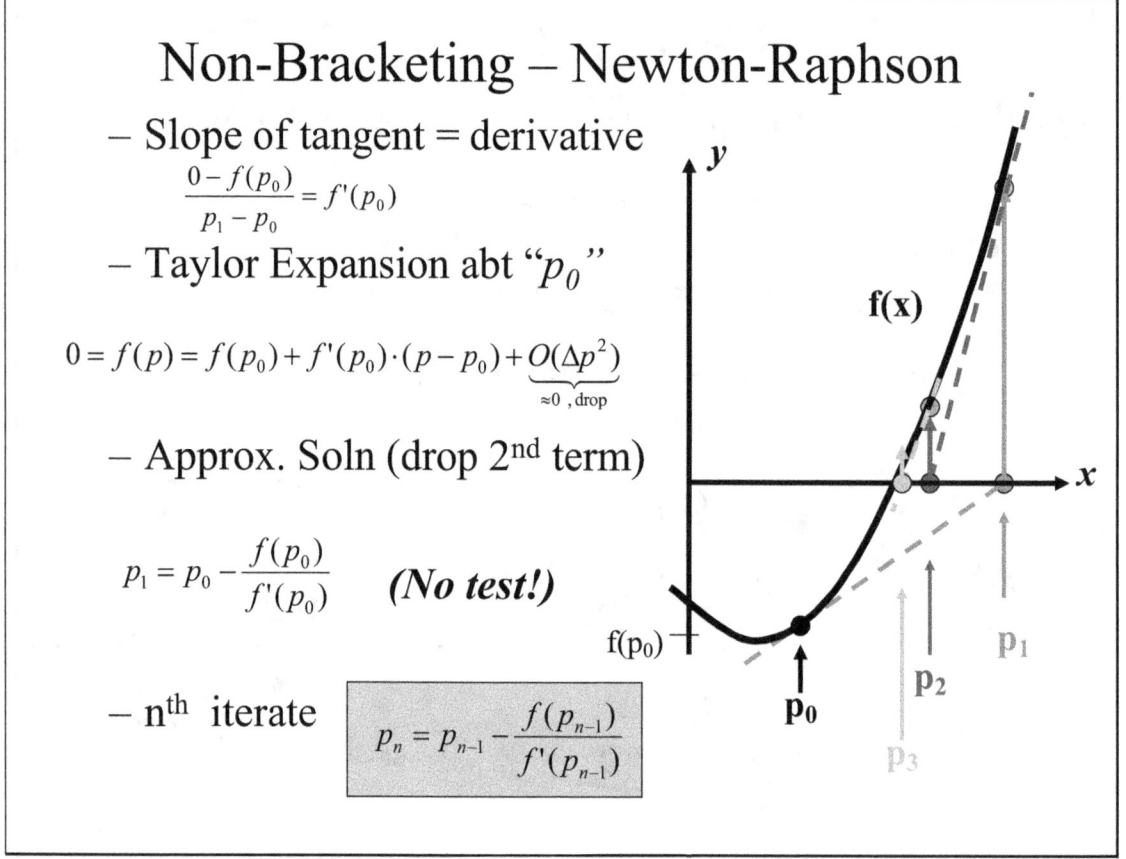

The Newton-Raphson method involves a single starting point p_0, but also requires the derivative at that point $f'(p_0)$. It is easily derived by letting the intersection of the geometrical tangent to the curve with the x-axis determine a new iterate p_1 and then equating the slope of this straight line segment to the derivative $f'(p_0)$. Alternately a Taylor series approximation about the true, but unknown, root p is made and an equality is obtained by dropping the truncation error terms and letting $p \rightarrow p_1$ a new estimate of the root

$$p_1 = p_0 - f(p_0) / f'(p_0).$$

Note that as p_{n-1} approaches p_{n-2}, the slope of the secant line between these points approximates the tangent to the curve at the point p_{n-1} and hence the secant method becomes identical to the N-R method in the calculus limit.

The two methods are quite distinct, however, because whereas the N-R method requires evaluation of the function and its "analytic" derivative at a single point, the secant method places no restriction on the proximity of the two points. Any thought of replacing the analytic derivative in N-R by a numerical approximation is quite problematic, since, as we have already discussed, obtaining a good numerical estimate for a derivative requires great care. Because the secant method does not place any restriction on the proximity of the two points it is a distinct alternative to the N-R method.

2.4.4 Problems With Newton-Raphson

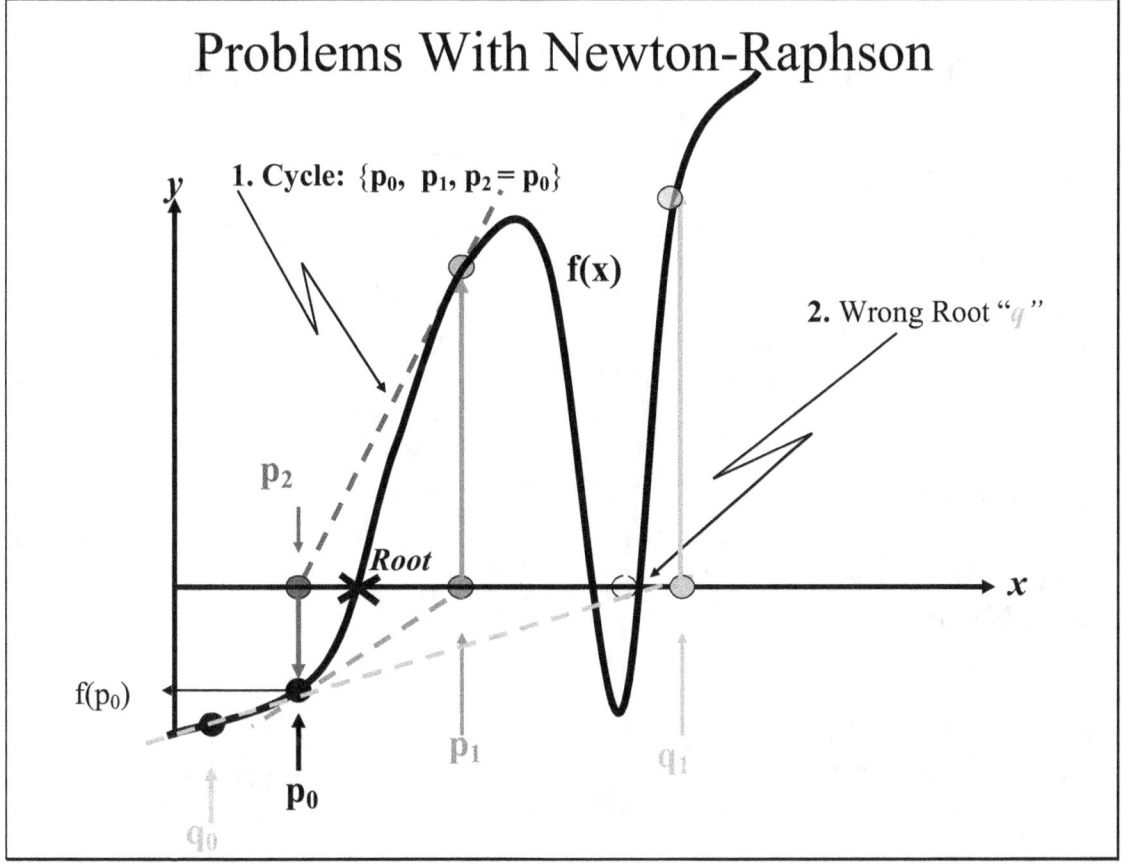

This slide illustrates two issues with the Newton-Raphson method.

Cycle: The desired root is shown as an **X** on the x-axis. Starting with the initial guess p_0 and the corresponding curve point $(p_0, f(p_0))$ we draw a tangent line (red dashed) and find that it intersects the x-axis at the new iterate p_1; following the arrow vertically up to the curve yields the point $(p_1, f(p_1))$. The tangent line (blue dashed) from this point intersects the x-axis at a new point p_2 (which is identical to p_0) and finally the vertical arrow down to the curve returns us to starting point $(p_0, f(p_0))$ completing the "cycle." Clearly this cycle continues endlessly except for minor perturbations caused by round off error and it never leads to the desired root **X**.

Wrong Root: Still searching for the same root **X** we start with a new initial guess q_0 and the curve point $(q_0, f(q_0))$ from which we project a new tangent line (green dashed) that intersects the x-axis at the new iterate q_1; again following the arrow vertically up to the curve yields the point $(q_1, f(q_1))$. Finally, the tangent line (yellow dashed) emanating from this point intersects the x-axis at a point q_2 which is converging on the wrong root.

2.5 *Root Multiplicity & Convergence Rate*

Root Multiplicity & Convergence Rate

NR algorithm

$$p_n = p_{n-1} - \frac{f(p_{n-1})}{f'(p_{n-1})}$$

Root "p" multiplicity "m"

$$f(x) = (x-p)^m \cdot q(x)$$

NOTE:Numerically"nearly equal" roots mimic "multiple roots"

$$f'(x) = m(x-p)^{m-1} \cdot q(x) + (x-p)^m \cdot q'(x)$$

$$\varepsilon_n \le \varepsilon_{n-1}\left(1 - \frac{1}{m}\right) + \varepsilon_{n-1}^2 \frac{q'_{n-1}}{m^2 q_{n-1}} + O(\varepsilon_{n-1}^3)$$

Quadratic convergence

$$\frac{\varepsilon_n}{\varepsilon_{n-1}^2} \le \frac{q'_{n-1}}{q_{n-1}} \qquad m = 1$$

Linear convergence for multiple roots

$$\Rightarrow \qquad \frac{\varepsilon_n}{\varepsilon_{n-1}} \le \left(1 - \frac{1}{m}\right) \qquad m \ne 1$$

NR using ratio function again yields Quadratic convergence!

$$\mu(x) \equiv \frac{f(x)}{f'(x)} \qquad p_n = p_{n-1} - \frac{\mu(p_{n-1})}{\mu'(p_{n-1})}$$

The polynomial $(x-1)^6$ has a six-fold multiple root at x=1 and is said to have a multiplicity m=6. The plot of this polynomial (see the polynomial cancellation example on Slide#1-8) shows the curve to be *very flat* near the root at x=1, as might be expected from the fact that the first five derivatives of this function are all zero at x=1.

Inasmuch as the N-R root finding algorithm divides f(x) by the first derivative f'(x) which is zero at x=1, we should also expect convergence issues near the multiple root at x=1; in fact the convergence of the algorithm ceases to be quadratic. This slide shows the mathematical reason for this slowing down of the convergence rate and gives a method for re-gaining the fast quadratic convergence in terms of a ratio function $\mu(x) = f(x) / f'(x)$.

The general form for a function with a root at x = p with multiplicity m is $f(x) = q(x)*(x-p)^m$, where q(x) is an arbitrary function *having no roots* at p. Evaluating this function and its derivative at x= p_{n-1} and using the definition of the error $\varepsilon_n = p-p_n$ we find the boxed equation for error at the nth iteration ε_n in terms of a linear term ε_{n-1} and a quadratic term ε_{n-1}^2.

For multiplicity m=1, the linear term vanishes, and we have quadratic convergence; for all other multiplicities 2, 3,... the linear term is the dominant term and the iteration sequence converges only linearly. If we instead apply N-R method to the ratio function $\mu(x) = f(x) / f'(x)$, we again obtain quadratic convergence. For example, taking $f(x)=(x-p)^3 q(x)$, then $f'(x)= [3(x-p)^2 q(x)+(x-p)^3 q'(x)]$, and the ratio $\mu(x) = (x-p)^3 q(x)/[3(x-p)^2 q(x)+(x-p)^3 q'(x)] = (x-p)^1 q(x)/[3q(x)+(x-p)^1 q'(x)] \rightarrow (x-p)^1/3$ for $x \rightarrow p$; thus the ratio is linear in (x-p) and hence NR regains quadratic convergence!

2.6 *Extensions of Newton-Raphson*

Extensions of Newton-Raphson

2nd Order Taylor Series:

$$0 = f(p) = f(p_n) + f'(p_n)\underbrace{(p - p_n)}_{\equiv \Delta p_n} + f''(p_n)\frac{(p - p_n)^2}{2} + error$$

$$\Delta p = p - p_n \cong \frac{-f_n' \pm \sqrt{f_n'^2 - 2f_n f_n''}}{f_n''}$$

$$p_{n+1} - p_n = \frac{-f_n' + \sqrt{f_n'^2 - 2f_n f_n''}}{f_n''} \cdot \left[\frac{-f_n' - \sqrt{f_n'^2 - 2f_n f_n''}}{-f_n' - \sqrt{f_n'^2 - 2f_n f_n''}}\right] = \frac{+2f_n f_n''}{f_n''[-f_n' - \sqrt{f_n'^2 - 2f_n f_n''}]}$$

Solving for p_{n+1} yields

$$p_{n+1} = p_n - \frac{2f_n}{f_n'\left[1 + \sqrt{1 - \dfrac{2f_n f_n''}{f_n'^2}}\right]} \qquad Cauchy$$

Expanding Square Root yields

$$p_{n+1} = p_n - \frac{f_n}{f_n'\left[1 - \dfrac{f_n f_n''}{2f_n'^2}\right]} \qquad Halley$$

Expanding the denominator yields

$$p_{n+1} = p_n - \frac{f_n}{f_n'}\left[1 + \dfrac{f_n f_n''}{2f_n'^2}\right] \qquad Chebyshev$$

If we make a Taylor expansion of the function f(x) about x = p to second order in stepsize h, we obtain several generalizations of the N-R method. In order to avoid subtracting nearly equal numbers in the solution for Δp, we rationalize the expression (just as we did for the Muller method) to find the Cauchy formula with a choice of two signs. Choosing the positive root (which yields the smallest Δp value), we obtain the Cauchy method; upon expanding the square root in the denominator, we obtain an approximation called the Halley method; further approximating by expanding the denominator of the Halley formula, we obtain the Tschebyshev method.

Note that these formulas will converge cubically, but require both first and second derivatives of f(x). They all reduce to the N-R formula if the second derivative vanishes or the term involving f, f', and f'' approaches zero. Some care must be taken in order to avoid singularities in the denominator.

2.6.1 Extensions of Newton-Raphson -2 dim

Extensions of Newton-Raphson - 2 dim

Simult. Zero of 2 Fcns: $\quad f_1(x,y)=0 \quad ; \quad f_2(x,y)=0$

2^d Taylor Series:	2^d Matrix-Vector Form:
$0=f_1(x,y)=f_1(x_0,y_0)+\dfrac{\partial f_1}{\partial x}\bigg\|_{(x_0,y_0)}\cdot(x-x_0)+\dfrac{\partial f_1}{\partial y}\bigg\|_{(x_0,y_0)}\cdot(y-y_0)+O(x^2)$ $0=f_2(x,y)=f_2(x_0,y_0)+\dfrac{\partial f_2}{\partial x}\bigg\|_{(x_0,y_0)}\cdot(x-x_0)+\dfrac{\partial f_2}{\partial y}\bigg\|_{(x_0,y_0)}\cdot(y-y_0)+O(x^2)$	$\bar{\bar{F}}\cdot\Delta\vec{x}=\vec{f}$ $\bar{\bar{F}}=\begin{bmatrix} f_{1x} & f_{1y} \\ f_{2x} & f_{2y} \end{bmatrix}$; $\Delta\vec{x}=\begin{bmatrix} x-x_0 \\ y-y_0 \end{bmatrix}$; $\vec{f}=\begin{bmatrix} -f_1 \\ -f_2 \end{bmatrix}$
2 Eqns; 2 Unknowns: $\dfrac{\partial f_1}{\partial x}\bigg\|_{(x_0,y_0)}\cdot(x-x_0)+\dfrac{\partial f_1}{\partial y}\bigg\|_{(x_0,y_0)}\cdot(y-y_0)\cong -f_1(x_0,y_0)$ $\dfrac{\partial f_2}{\partial x}\bigg\|_{(x_0,y_0)}\cdot(x-x_0)+\dfrac{\partial f_2}{\partial y}\bigg\|_{(x_0,y_0)}\cdot(y-y_0)\cong -f_2(x_0,y_0)$	Soln: $\quad \Delta\vec{x}=\bar{\bar{F}}^{-1}\cdot\vec{f} \quad \|J\|=\|\det F\|$ $\begin{bmatrix} x-x_0 \\ y-y_0 \end{bmatrix}=\dfrac{1}{\|J\|}\cdot\begin{bmatrix} f_{2y} & -f_{1y} \\ -f_{2x} & f_{1x} \end{bmatrix}\cdot\begin{bmatrix} -f_1 \\ -f_2 \end{bmatrix}$ $=\dfrac{1}{\|J\|}\cdot\begin{bmatrix} (-f_1 f_{2y}+f_2 f_{1y}) \\ (f_{2x}f_1 - f_{1x}f_2) \end{bmatrix}$

Iterative Solution:

$$x_1 = x_0 - \|J\|^{-1}\cdot(f_1 f_{2y} - f_2 f_{1y})\big|_0$$

$$y_1 = y_0 - \|J\|^{-1}\cdot(-f_{2x}f_1 + f_{1x}f_2)\big|_0$$

$$\|J\|=\det\begin{bmatrix} f_{1x} & f_{1y} \\ f_{2x} & f_{2y} \end{bmatrix}$$

The N-R method can be extended to multiple dimensions. Here we show the N-R method applied to finding x, y pairs which simultaneously provide zeros for 2 functions $f_1(x, y) = f_2(x, y) = 0$. In the upper panel, the two-dimensional (2^d) Taylor series in x and y about the point (x_0, y_0) is written down to first order in small quantities for both functions; a simpler matrix-vector form is written on the right hand side.

This leads directly to two equations in two unknowns (middle panel) and to the more transparent 2^d matrix-vector form on the right. Upon dropping higher order terms and replacing x by x_1 and y by y_1 we arrive at an update equation (bottom panel) for x_1 and y_1 given the initial guesses x_0 and y_0. This represents the first update in the 2^d Newton-Raphson iteration sequence

$$\{(x_0,y_0), (x_1,y_1),\ldots, (x_{n-1},y_{n-1}), (x_n,y_n)\}$$

The quantity $\|J\|$ is the absolute value of the determinant of Jacobian partial derivatives matrix **K** evaluated at (x_0, y_0). Note that the inverse of the 2-dim matrix K is obtained by (i) swapping the diagonals, (ii) changing the sign of the off-diagonals, and (iii) dividing by the scalar $\|J\| = \|\det K\|$.

2.7 *Convergence Rates Comparison*

Convergence Rates Comparison

$$\varepsilon_n \equiv p - p_n$$

$$\varepsilon_n = C_L \cdot \left(\varepsilon_{n-1}\right)^1 \qquad \text{Linear Convergence}$$

$$\varepsilon_n = C_Q \cdot \left(\varepsilon_{n-1}\right)^2 \qquad \text{Quadratic Convergence}$$

$$\varepsilon_n = C_C \cdot \left(\varepsilon_{n-1}\right)^3 \qquad \text{Cubic Convergence}$$

$\varepsilon_n = C_\alpha \cdot \left(\varepsilon_{n-1}\right)^\alpha$	Bisection	Muller	Secant	N-R	Cauchy
α	1.0	1.84	1.62	2.0	3.0
Robustness	Excellent	Good	OK	No	No

This short table compares the rates of convergence and robustness for some common methods we have discussed. The bisection and false position methods are robust in the sense that they will always find a root, but only have linear convergence. The Secant and Muller methods have super-linear convergence and are also fairly robust, while the NR and Cauchy methods have quadratic and cubic convergence respectively, but require analytic derivatives, and do not always find the desired root.

2.8 *Fixed Point Methods*

Fixed Point Methods

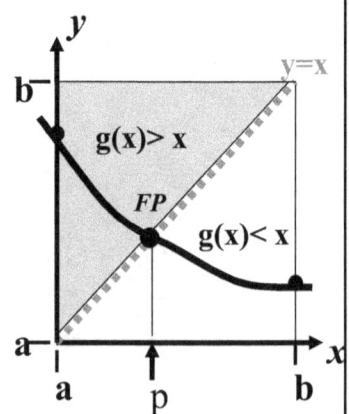

- **Decomposition:** $f(x) = g(x) - x$
- **Equivalence:** $f(x) = 0 \Leftrightarrow x = g(x)$
- **$x = p$, a fixed point:** $g(p) = p$

- **Decomposition non-unique:**

$$f(x) = x^3 - 4x^2 + 3x + 1 = 0$$

$$g_1(x) = \left(4x^2 - 3x - 1\right)^{1/3}$$
$$g_2(x) = \sqrt{1 + 3x + x^3}\,/2$$
$$g_3(x) = 4 - \frac{3}{x} - \frac{1}{x^2}$$

We have discussed methods for finding the roots of a function f(x); if the function is re-written as
$$f(x) = g(x) - x,$$
then zeros of f(x) correspond to points at which
$$x = g(x).$$
Such points are called "fixed points" because re-applying the function g() to this latter equality yields g(x) = g(g(x)) which is again equal to "x". This "functional iteration" process will be discussed in more detail on the next slide.

Graphically a fixed point occurs at the intersection of the curve y = g(x) with the line y = x. If we consider a root in the interval [a,b] we see that the line y = x divides the "square box" x ε [a,b] y ε [a,b] into points above the (red) line where g(x)>x or f(x)>0 and points below the (red) line where g(x) < x or f(x)<0 . Thus, it is clear that if a fixed point (FP) of g(x) exists *i.e.,* p = g(p), then
$$f(p) = g(p) - p = 0,$$
so that x = p is a root of the original function f(x) in the interval [a,b].

The functional decomposition of f(x) = 0 into the form g(x) - x = 0 can be accomplished in many ways as shown by the example on this slide. The key is to determine under what circumstances g(x) has a "unique fixed point" in the interval [a,b]. The Fixed Point Theorem states conditions under which this is guaranteed.

2.8.1 Fixed Point Sequence

Fixed Point Sequence

Functional iteration:

$$p_n = g(\cdots g(g(p_0))) = [g]^n (p_0)$$

$$p_1 = g(p_0)$$

$$p_2 = g(p_1)$$

$$\vdots$$

$$p_n = g(p_{n-1}) \; ;$$

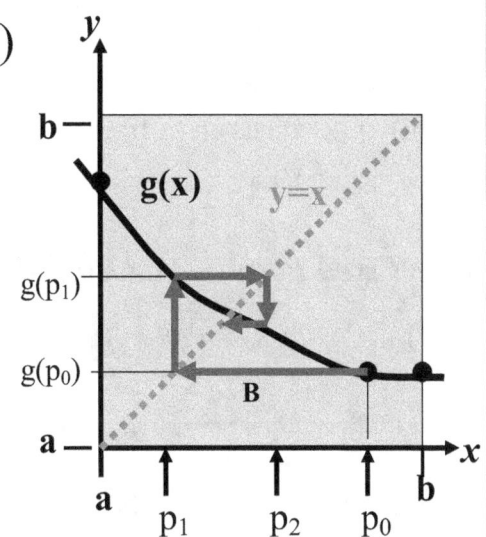

The fixed point method is also known as "functional iteration" because of the way we repeatedly apply the function $g(x)$ to the previous iterate. Thus, starting from the initial guess p_0, we obtain the 1st iterate $p_1 = g(p_0)$, the second iterate $p_2 = g(p_1) = g(g(p_0))$, ... , *etc.* .

When the function $g(x)$ and its derivative satisfy certain conditions this sequence converges to a fixed point which is the intersection of the straight line $y=x$ and the curve $g(x)$. The blue (B) spiraling rectangular path shows the graphical steps corresponding to the functional iteration.

Starting at p_0 we go **vertically up** to the curve value $g(p_0)$ and then left **horizontally** towards the dashed straight line ($y=x$) to determine the iterate $p_1 = g(p_0)$; next **vertically up** to the curve $g(p_1)$ and then right **horizontally** towards the straight line again to determine the iterate $p_2 = g(p_1)$; .... This graph spirals in towards the fixed point because this particular choice of $g(x)$ satisfies some special properties described by the Fixed Point Theorem discussed on the next slide.

2.8.2 Fixed Point Theorem & Error Bounds

Fixed Point Theorem & Error Bounds

1. Fixed Point Theorem

If

   i) $g(x)$ continuous on $[a,b]$

   ii) $g([a,b]) \to [a,b]$ "maps"

   iii) $g'(x)$ exists on (a,b)

     & $|g'(x)| \le k < 1$

Then

   – $\{p_n\}$ converges to **unique fixed pt**

   – independent of initial guess "p_0"

   – *sufficient conditions for fixed pt*

   – *not necessary!*

2. Fixed Point Error Bounds

  i) $|p_n - p| \le k^n \cdot |p_0 - p| \le k^n \cdot \max\{p_0 - a, b - p_0\}$

  ii) $|p_n - p| \le \dfrac{k^n}{1-k} \cdot |p_0 - p_1|$

  where, $k \equiv \max|g'(x)|$

For given error ε_{TOL}, Estimate #iterations n
$\quad \dfrac{k^n}{1-k} \cdot |p_0 - p_1| = \varepsilon_{TOL} \;\Rightarrow\; n = \dfrac{\log\left[\varepsilon_{TOL}(1-k) \Big/ |p_0 - p_1|\right]}{\log k}$

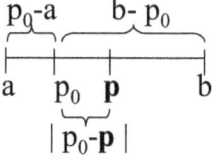

one of these two intervals will be larger than $|p_0\text{-}p|$

$p_0\text{-}a \qquad b\text{-}p_0$

$a \quad p_0 \quad p \qquad b$

$|\,p_0\text{-}p\,|$

The Fixed Point Theorem states the conditions under which we can expect functional iteration to converge to a unique fixed point. The following three conditions must be satisfied:
i) $g(x)$ is continuous on $[a,b]$,
ii) box maps $g([a,b])$ to $[a,b]$, and
iii) the derivative $g'(x)$ exists on (a,b) and its absolute value is always less than 1.

If these conditions hold, then we are guaranteed a *unique fixed point* and convergence of the iteration sequence p_n for *any initial guess* p_0 within the "vetted" region $[a,b]$.
It is important to note that these are **sufficient conditions** for the existence of a unique fixed point, but they are not necessary conditions. That is, a fixed point may very well exist even if these conditions do not hold; but we are absolutely guaranteed convergence to a unique fixed point if the conditions are true.
We display two fixed point error bounds for the nth iterate p_n . The second one is the most useful since it only depends upon the initial guess p_0, the first iterate $p_1 = g(p_0)$, and an upper bound of the derivative over the interval $[a,b]$ expressed in terms of $k = \max| g'[a,b]|$. From this second formula, we can estimate the number of iterations necessary to achieve a given error tolerance, $\varepsilon_n = |p - p_n| < 10^{-6}$ by solving (ii) directly for n.

2.8.3 Fixed Point Trajectories

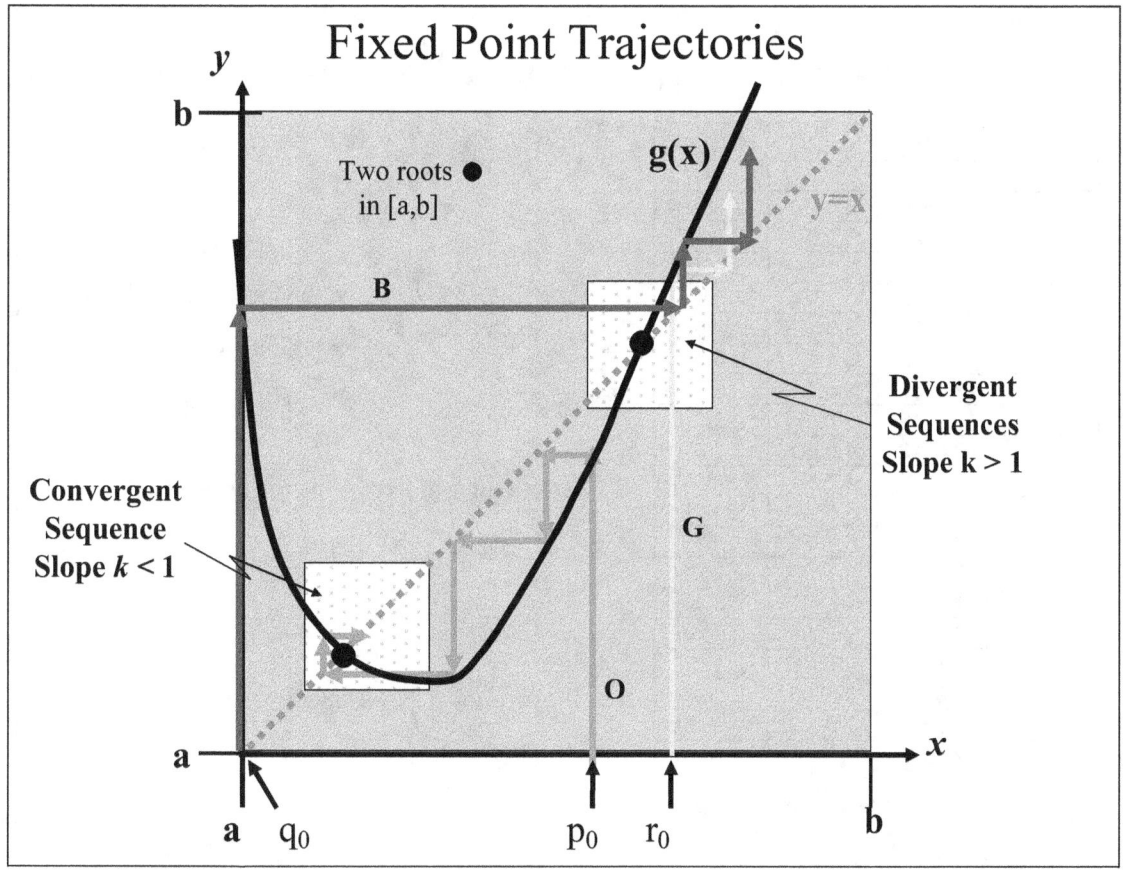

The function g(x) has two fixed points in the box [a, b] as evidenced by its two intersection with the red dashed line y = x; however, the fixed point theorem does not apply in this case because g(x) does not satisfy the "map on box" condition g([a, b]) ➔ [a,b] and moreover, the derivative |g′([a,b])| is not < 1 over the entire interval [a, b].

Even though the FP theorem does not apply, an initial guess of p_0 follows the orange (O) line and does indeed converge to the fixed point in the lower left region. However, arbitrary points in [a,b] such as r_0 (green (G) line sequence) and q_0 (blue (B) line sequence) clearly do not converge to a fixed point, but instead both diverge to points outside the interval [a,b].

Reducing the box interval as shown in the lower left portion of the figure, yields a new box region in which the function g(x) is seen to satisfy all conditions of the FP theorem and hence we can start anywhere within that interval and be assured convergence to the unique fixed point in the smaller box. On the other hand, for the box in the upper right portion of the figure we see that g(x) does not map the box region and, moreover, its derivative exceeds 1 and the functional iteration diverges away from that fixed point.

2.8.4 Fixed Point Iterates Example

Fixed Point Iterates Example

k	g1(x)	g2(x)	g3(x)
0	3	3	3
1	2.9625	3.0414	2.8889
2	2.9325	3.0926	2.8417
3	2.9084	3.1566	2.8205
4	2.8889	3.2374	2.8106
5	2.8731	3.3407	2.806
6	2.8602	3.4751	2.8039
7	2.8497	3.6535	2.8029
8	2.8412	3.8964	2.8024
9	2.8342	4.2381	2.8021
10	2.8284	4.7391	2.802
11	2.8237	5.5147	2.802
12	2.8199	6.8055	2.802
13	2.8167	9.1736	2.8019
14	2.8141	14.147	2.8019
15	2.8119	26.808	2.8019
16	2.8102	69.546	2.8019
17	2.8087	290.08	2.8019
18	2.8075	2470.3	2.8019
19	2.8065	61390	2.8019

Diverges !

$$f(x) = x^3 - 4x^2 + 3x + 1 = 0$$

Root: $x \, \varepsilon \, [2.5, 4]$

FP Thm

$g_1(x) = (4x^2 - 3x - 1)^{1/3}$ → 2.5458 3.7084 Maps

$g_2(x) = \sqrt{1 + 3x + x^3} / 2$ → 2.4559 4.3875 NO

$g_3(x) = 4 - \dfrac{3}{x} - \dfrac{1}{x^2}$ → 2.6400 3.1875 Maps

Estimated # of iterations for error $=10^{-4}$

$$k = \max|g_3'(x)| = \max_{[2.5,4]}(3/x^2 + 2/x^3) = .608$$

$$\varepsilon_n \cong \frac{k^n}{1-k} \cdot |p_0 - p_1|$$

$$10^{-4} \cong \frac{.608^n}{1-.608} \cdot |3 - 2.8889|$$

$$\Rightarrow n = 16$$

This example problem asks us to find a root of the function f(x) in the interval [2.5, 4] using the fixed point method. Now the function f(x) = 0 can be "manipulated" to solve for "x" equal to g(x) in a multitude of ways, some of which are displayed as $g_1(x)$, $g_2(x)$, and $g_3(x)$ in this slide. (See MatLab script on Slide#5-4)

WARNING: None of these "g(x)"s are obtained by simply adding a function of x to each side of the equation f(x) = 0 to obtain, for example, $x^2 = f(x) + x^2 = g(x)$; such a procedure may occasionally yield a fixed point root, but most of the time does not produce anything useful!!

Instead, we separate one term from the expression f(x) = 0, e.g., $x^3 = 4x^3 - 3x-1$ and then solve for "x" by taking the cube root to obtain $g_1(x)$. The other "g(x)"s are obtained in a similar manner. There is no guarantee that a particular g(x) will work; vetting each one individually, using the FP theorem, shows that $g_1(x)$ and $g_3(x)$ map within the box [2.5, 4] while $g_2(x)$ does not. You must also verify that the derivatives |g'(x)| <1 everywhere in the interval [2.5, 4]. I have not done this on the slide, but you should!! Checking the endpoints is sufficient only if the function g(x) and its derivative g'(x) are monotonic increasing or decreasing; otherwise you must check for extremum values to make sure that the function does not "pop up" somewhere.

The results of the table show that iterates for $g_1(x)$ and $g_3(x)$ both converge, with the latter being faster; $g_2(x)$ might have converged even though FP Theorem is not satisfied, but it in fact diverges.

Note the number of iterations required for $g_3(x)$ to converge with $\varepsilon < 10^{-4}$ is estimated to be n=16 compared with the table convergence at n=13; this is OK since the formula is an approximation that just gives an *upper bound* for the required number of iterations.

2.9 *Root Finding Methods Compared: 5 sd Good* x_0

Root Finding Methods Compared: 5 sd Good x_0

$$P(x) = x^3 - 2x^2 - 5 \qquad x_0 = 2, \ x_1 = 2.5, \ x_2 = 3.0 \qquad g_1(x) = (2x^2 + 5)^{1/3}$$
$$g_2(x) = 2 + 5/x^2$$

		Bisection		FP				
Secant	NwtnRaphs	an	bn	pn	g1	g2	Cheb	Muller
2	2	2	3	0	2	2	2	2
2.5	3.25	2.5	3	2.5	2.3513	3.25	1.6875	2.5
2.8	2.811	2.5	2.75	2.75	2.5229	2.4734	-13.4595	3
2.6787	2.698	2.625	2.75	2.625	2.6076	2.8173	-7.1971	2.687
2.69	2.6907	2.6875	2.75	2.6875	2.6495	2.6299	-3.7085	2.6906
2.6907	2.6906	2.6875	2.7188	2.7188	2.6702	2.7229	-1.7011	2.6906
2.6906	2.6906	2.6875	2.7031	2.7031	2.6805	2.6744	-0.2145	2.6906
2.6906	2.6906	2.6875	2.6953	2.6953	2.6856	2.6991	74.534	2.6906
2.6906	2.6906	2.6875	2.6914	2.6914	2.6882	2.6863	41.7107	NaN
2.6906	2.6906	2.6895	2.6914	2.6895	2.6894	2.6929	23.4816	NaN
NaN	2.6906	2.6904	2.6914	2.6904	2.69	2.6895	13.3671	NaN
NaN	2.6906	2.6904	2.6909	2.6909	2.6903	2.6912	7.7768	NaN
NaN	2.6906	2.6904	2.6907	2.6907	2.6905	2.6903	4.7414	NaN
NaN	2.6906	2.6906	2.6907	2.6906	2.6906	2.6908	3.2291	NaN
NaN	2.6906	2.6906	2.6907	2.6906	2.6906	2.6906	2.7277	NaN
NaN	2.6906	2.6906	2.6907	2.6906	2.6906	2.6907	2.6907	NaN
NaN	2.6906	2.6906	2.6907	2.6907	2.6906	2.6906	2.6906	NaN
NaN	2.6906	2.6906	2.6907	2.6907	2.6906	2.6907	2.6906	NaN
NaN	2.6906	2.6906	2.6907	2.6906	2.6906	2.6906	2.6906	NaN
NaN	2.6906	2.6906	2.6906	2.6906	2.6906	2.6907	2.6906	NaN
NaN	2.6906	2.6906	2.6906	2.6906	2.6906	2.6906	2.6906	NaN
NaN	2.6906	2.6906	2.6906	2.6906	2.6906	2.6906	2.6906	NaN

Here is comparison of Secant, Newton-Raphson, bisection, fixed point Tschebyshev (Cheb), and Muller algorithms for $P(x) = x^3 - 2x^2 - 5$ using a=2,b=3 for Bisection and Secant methods, $x_0=2$, $x_1=2.5$, $x_2=3.0$ for Muller and using $x_0=2$ for Newton-Raphson, Tschebyshev, and the two fixed point methods, $g_1(x)$ and $g_2(x)$. (See MatLab script on Slides#5-13 to 5-15).

Comparing the results, we see that Bisection is the slowest and that, Secant, N-R, and Muller converge with 5 sd within ~6 iterations, while Bisection takes ~20 iterations and the fixed point methods both take ~10 iterations. The Tschebyshev method has some initial random looking jumps, but finally settles down and converges in ~15 iterations.

2.9.1 Root Finding Methods Compared: 5 sd Bad x_0

Root Finding Methods Compared: 5 sd Bad x_0

$$P(x) = x^3 - 2x^2 - 5$$

Poor initial guess
$x_0 = -12, x_1 = 2.5, x_2 = 3.0$

$g_1(x) = (2x^2 + 5)^{1/3}$

$g_2(x) = 2 + 5/x^2$

| Secant | NwtnRaphs | Bisection | | FP | | | | |
		an	bn	pn	g1	g2	Cheb	Muller
-12	-12	-12	3	0	-12	-12	-12	-12
2.5	-7.7896	-4.5	3	-4.5	6.6419	2.0347	-6.3861	2.5
2.5135	-4.9799	-0.75	3	-0.75	4.5344	3.2077	-3.2521	3
2.7125	-3.0916	1.125	3	1.125	3.5862	2.4859	-1.4174	2.6256
2.6884	-1.784	2.0625	3	2.0625	3.1319	2.8091	0.1472	2.6903
2.6906	-0.7624	2.5313	3	2.5313	2.9091	2.6336	-285.1595	2.6906
2.6906	0.6156	2.5313	2.7656	2.7656	2.7989	2.7209	-158.1272	2.6906
2.6906	-3.5524	2.6484	2.7656	2.6484	2.7443	2.6754	-87.5548	2.6906
2.6906	-2.1106	2.6484	2.707	2.707	2.7172	2.6985	-48.3499	2.6906
2.6906	-1.0416	2.6777	2.707	2.6777	2.7038	2.6866	-26.5725	NaN
NaN	0.0768	2.6777	2.6924	2.6924	2.6972	2.6927	-14.4784	NaN
NaN	-17.2377	2.6851	2.6924	2.6851	2.6939	2.6896	-7.7631	NaN
NaN	-11.2803	2.6887	2.6924	2.6887	2.6923	2.6912	-4.0256	NaN
NaN	-7.3098	2.6906	2.6924	2.6906	2.6914	2.6904	-1.8924	NaN
NaN	-4.6589	2.6906	2.6915	2.6915	2.691	2.6908	-0.4098	NaN
NaN	-2.8734	2.6906	2.691	2.691	2.6908	2.6906	11.6959	NaN
NaN	-1.626	2.6906	2.6908	2.6908	2.6907	2.6907	6.86	NaN
NaN	-0.6155	2.6906	2.6907	2.6907	2.6907	2.6906	4.2612	NaN
NaN	1.0494	2.6906	2.6907	2.6906	2.6907	2.6907	3.0304	NaN
NaN	-5.7145	2.6906	2.6907	2.6906	2.6907	2.6906	2.7027	NaN
NaN	-3.5881	2.6906	2.6907	2.6907	2.6907	2.6907	2.6906	NaN
NaN	-2.1357	2.6906	2.6907	2.6906	2.6907	2.6906	2.6906	NaN

Here is the same (but now with $x_0=-12$) comparison as the previous slide of Secant, Newton-Raphson, bisection, fixed point Tschebyshev (Cheb), and Muller algorithms for the polynomial $P(x) = x^3-2x^2-5$ using **a=-12,b=3** for Bisection and Secant methods, $x_0=-12, x_1=2.5$, $x_2=3.0$ for Muller and using $x_0=-12$ for Newton-Raphson, Tschebyshev and the two fixed point methods $g_1(x)$ and $g_2(x)$. (See MatLab script on Slides#5-13 to 5-15).

Comparing the results, we see that Bisection is again the slowest taking ~20 iterations, and that Secant and Muller still converge in ~6 iterations, while the fixed point methods now both take ~20 iterations; the Tschebyshev method again has some initial random looking jumps, but finally settles down and converges now in ~20 iterations. However, the N-R iterates do not converge at all showing its extreme sensitivity to the initial guess.

2.9.2 Non-Bracketing Root Finding Methods: 16 sd Good x_0

Non-Bracketing Root Finding Methods: 16 sd Good x_0

$P(x) = x^3 - 2x^2 - 5$ $x_0 = 2$, $x_1 = 2.5$, $x_2 = 3.0$

Secant	NwtnRaphs	Cheb	Cauchy	Muller
2.00000000000000	2.00000000000000	2.00000000000000	2.00000000000000	2.00000000000000
2.50000000000000	3.25000000000000	1.68750000000000	2.72474487139158	2.50000000000000
2.80000000000000	2.81103678929765	-13.45951491311820	2.69064382855482	3.00000000000000
2.67874165872259	2.69798950246852	-7.19714754209142	2.69064744802861	2.68697019943163
2.68995254827652	2.69067715286036	-3.70846849940176	2.69064744802861	2.69062775986006
2.69065205573538	2.69064744851761	-1.70109276848371	2.69064744802861	2.69064744598462
2.69064744625363	2.69064744802861	-0.21450775304470	2.69064744802861	2.69064744802861
2.69064744802860	2.69064744802861	74.53403584248080	2.69064744802861	2.69064744802861
2.69064744802861	2.69064744802861	41.71068071640960	2.69064744802861	NaN
2.69064744802861	2.69064744802861	23.48157263760990	2.69064744802861	NaN
NaN	2.69064744802861	13.36708302668560	2.69064744802861	NaN
NaN	2.69064744802861	7.77682788354006	2.69064744802861	NaN
NaN	2.69064744802861	4.74140010853614	2.69064744802861	NaN
NaN	2.69064744802861	3.22913156117581	2.69064744802861	NaN
NaN	2.69064744802861	2.72769532915411	2.69064744802861	NaN
NaN	2.69064744802861	2.69067238018500	2.69064744802861	NaN
NaN	2.69064744802861	2.69064744802862	2.69064744802861	NaN
NaN	2.69064744802861	2.69064744802861	2.69064744802861	NaN
NaN	2.69064744802861	2.69064744802861	2.69064744802861	NaN
NaN	2.69064744802861	2.69064744802861	2.69064744802861	NaN
NaN	2.69064744802861	2.69064744802861	2.69064744802861	NaN

Here is full 16 digit comparison of the rapidly converging methods Secant, Newton-Raphson, Tschebyshev (Cheb), Cauchy, and Muller algorithms for $P(x) = x^3 - 2x^2 - 5$ using a=2, b=3.0 for Secant, $x_0=2$, $x_1=2.5$, $x_2=3.0$ for Muller, and $x_0=2$ for Newton-Raphson, Tschebyshev, and Cauchy methods. (See MatLab script on Slides#5-13 to 5-15).

Comparing the results for convergence to 16 digits, Secant takes ~20 iterations, N-R ~6 iterations, Tschebyshev (Cheb) ~15 iterations, Cauchy ~3 iterations, and Muller ~7 iterations.

2.9.3 Non-Bracketing Root Finding Methods: 16 sd Bad x_0

Non-Bracketing Root Finding Methods: 16 sd Bad x_0

Poor initial guess

$P(x) = x^3 - 2x^2 - 5$ $x_0 = -12$, $x_1 = 2.5$, $x_2 = 3.0$ **Table Multiplier = 100**

Secant	NwtnRaphs	Cheb	Cauchy	Muller
-0.120000000000000	-0.120000000000000	-0.120000000000000	-0.120000000000000	-0.120000000000000
0.025000000000000	-0.077895833333333	-0.063861476598669	-0.056842105263158 - 0.036462328299788i	0.025000000000000
0.025134649910233	-0.049798584452771	-0.032521476416542	-0.015083323035701 - 0.036261175486087i	0.030000000000000
0.027124829179849	-0.030916032444458	-0.014173784536050	-0.014132998322000 - 0.012817705950019i	0.026255632574199
0.026883677239888	-0.017839761439939	0.001471610403407	-0.004188123237976 - 0.013983188041101i	0.026902991509666
0.026906200693761	-0.007624487337478	-2.851595004250960	-0.003451949954651 - 0.013186574863580i	0.026906468091475
0.026906474826508	0.006155640592653	-1.581271653083980	-0.003453237240146 - 0.013187267795712i	0.026906474480299
0.026906474480281	-0.035523678943520	-0.875548045824951	-0.003453237240143 - 0.013187267795713i	0.026906474480286
0.026906474480286	-0.021106393114960	-0.483499007032844	-0.003453237240143 - 0.013187267795713i	0.026906474480286
0.026906474480286	-0.010416192639613	-0.265725352328831	-0.003453237240143 - 0.013187267795713i	NaN
NaN	0.000767788701968	-0.144783601994282	-0.003453237240143 - 0.013187267795713i	NaN
NaN	-0.172376971334400	-0.077630741942301	-0.003453237240143 - 0.013187267795713i	NaN
NaN	-0.112803242730170	-0.040256075324703	-0.003453237240143 - 0.013187267795713i	NaN
NaN	-0.073097705920137	-0.018923929099849	-0.003453237240143 - 0.013187267795713i	NaN
NaN	-0.046588593807966	-0.004098168566959	-0.003453237240143 - 0.013187267795713i	NaN
NaN	-0.028734298456662	0.116959061319846	-0.003453237240143 - 0.013187267795713i	NaN
NaN	-0.016259513716785	0.068600423201501	-0.003453237240143 - 0.013187267795713i	NaN
NaN	-0.006154883467951	0.042611904437275	-0.003453237240143 - 0.013187267795713i	NaN
NaN	0.010493524045715	0.030304023665583	-0.003453237240143 - 0.013187267795713i	NaN
NaN	-0.057144913042409	0.027027088735055	-0.003453237240143 - 0.013187267795713i	NaN
NaN	-0.035880970238836	0.026906483464633	-0.003453237240143 - 0.013187267795713i	NaN
NaN	-0.021356649012103	0.026906474480286	-0.003453237240143 - 0.013187267795713i	NaN

$x_G = -.3453237240143 - 1.318726779571\,\mathbf{i}$; $x_G^* = -.3453237240143 + 1.318726779571\,\mathbf{i}$

$(x-x_G)(x-x_G^*)(x-x_2) = x^3 + x^2(-x_G - x_G^* - x_2) + x(x_G x_G^* + x_2 x_G + x_2 x_G^*) - x_G x_G^* x_2$

Prod. $-x_G x_G^* x_2 = -5$ → $x_2 = 5 / x_G x_G^*$ or $x_2 = 2.69064748802861$

Here is full 16 digit comparison of the rapidly converging methods, but with $x_0 = -12$ for the lower value both Secant and Muller, and $x_0 = -12$ for the starting values of Newton-Raphson, Tschebyshev (Cheb), Cauchy again for $P(x) = x^3 - 2x^2 - 5$. (See MatLab script on Slides#5-13 to 5-15).

Comparing the results (noting the table multiplier of 100) for convergence to 16 digits, Secant and Muller are the "steady troopers" taking ~8 iterations, Tschebyshev again goes through its "gyrations" and eventually converges ~20 iterations, and Cauchy converges to an **imaginary root** ~6 iterations. The interesting thing about the Cauchy method besides its steady fast convergence is that when started with a poor initial guess at $x_0 = -12$, instead of diverging like N-R, it finds the complex root with the added "bonus" of obtaining a pair of roots (complex conjugates) automatically. For this cubic polynomial the product of the roots is given by the constant term 5 and hence we can obtain the final real root by dividing the product 5 by $x \cdot x^*$ as shown at the bottom of the slide. Thus all three roots are obtained at once!

2.10 *Polynomial Roots*

Polynomial Roots

- m^{th} degree polynomial

$$P_m(x) = a_0 + a_1 \cdot x^1 + a_2 \cdot x^2 + \cdots + a_m \cdot x^m$$

- Synthetic Division: Quotient and Remainder

$$\frac{P_m(x)}{(x - x_0)} = \underbrace{b_1 + b_2 \cdot x^1 + b_3 \cdot x^2 + \cdots + b_m \cdot x^{m-1}}_{\equiv Q_{m-1}(x)} + \underbrace{\frac{b_0}{(x - x_0)}}_{R(x)}$$

- Evaluating the Polynomial at x_0

$$P_m(x)\Big|_{x=x_0} = (x - x_0) \cdot Q_{m-1}(x) + b_0\Big|_{x=x_0} = b_0$$

- Evaluating Polynomial derivative at x_0

$$P_m'(x)\Big|_{x=x_0} = (x - x_0) \cdot Q'_{m-1}(x) + Q_{m-1}(x)\Big|_{x=x_0} = Q_{m-1}(x_0)$$

Formally dividing an mth degree polynomial $P_m(x)$ by $(x-x_0)$ yields a polynomial of degree m-1 called the quotient $Q_{m-1}(x)$ and a remainder fraction $R(x) = b_0/(x-x_0)$. If we formally multiply back by $(x-x_0)$ we obtain an expression for the original polynomial and its derivative as

$$P_m(x) = (x-x_0) * Q_{m-1}(x) + b_0, \text{ and}$$
$$P'_m(x) = (x-x_0) * Q'_{m-1}(x) + Q_{m-1}(x).$$

Evaluating these two expressions at $x=x_0$, provides the basis for a method called "synthetic division" (see the next slide) which allows us to evaluate both the function and its first derivative at the point x_0. The fact that synthetic division (Horner's method) effectively nests polynomials makes it the most efficient numerical method to perform N-R iterations to find a zero of a polynomial $P_m(x)$.

Explicitly performing the multiplication $P_m(x) = (x-x_0) * Q_{m-1}(x) + b_0$ and comparing terms with the same exponent on the two sides yields a nested recursion relation between the known coefficients a_k of the original polynomial and the unknown coefficients b_k of the quotient polynomial. Solving the recursion equations sequentially for the "b_k"s gives an evaluation of the polynomial $P_m(x_0) = b_0$ as well as the coefficients of the quotient polynomial $Q_{m-1}(x)$. A second synthetic division on the quotient polynomial $Q_{m-1}(x) = (x-x_0) * R_{m-2}(x) + c_0$ yields an analogous result for evaluation of the polynomial derivative, $P'_m(x_0) = Q_{m-1}(x_0) = c_0$.

The next slide gives the recursion relation and explicitly shows the nesting for a simple example, while the following two slides give the general tableau and an explicit example. The synthetic division recursion relation and the tableau associated with it constitute Horner's method.

2.10.1   Horner's Method - Recursions

Horner's Method - Recursions

- ## Recursion relations for b's given a's

$$b_m = a_m$$

$$b_k = a_k + x_0 \cdot b_{k+1} \quad ; \quad k = m-1,\, m-2,\cdots,0$$

- ## Nesting (case m=4)

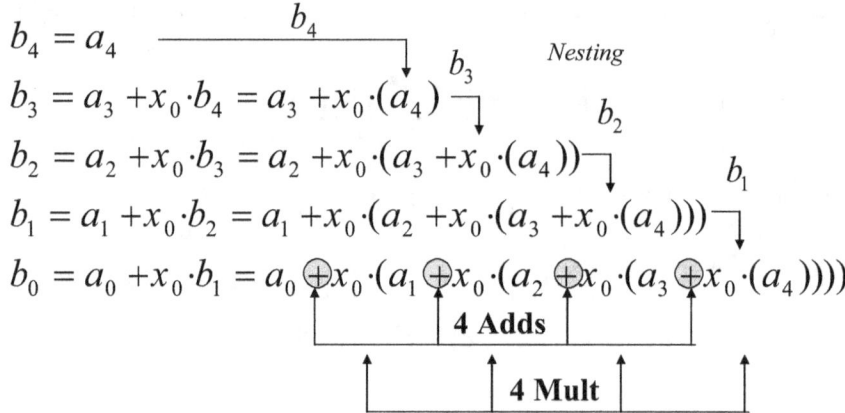

$$b_4 = a_4$$

$$b_3 = a_3 + x_0 \cdot b_4 = a_3 + x_0 \cdot (a_4)$$

$$b_2 = a_2 + x_0 \cdot b_3 = a_2 + x_0 \cdot (a_3 + x_0 \cdot (a_4))$$

$$b_1 = a_1 + x_0 \cdot b_2 = a_1 + x_0 \cdot (a_2 + x_0 \cdot (a_3 + x_0 \cdot (a_4)))$$

$$b_0 = a_0 + x_0 \cdot b_1 = a_0 \oplus x_0 \cdot (a_1 \oplus x_0 \cdot (a_2 \oplus x_0 \cdot (a_3 \oplus x_0 \cdot (a_4))))$$

4 Adds

4 Mult

Performing the multiplication discussed on the last slide yields the recursion relations) between the b's and a's. Note that the first relation identifies b_m with a_m, and proceeds "backwards" to the lower index b_{m-1} in terms of a_{m-1} and the now known value of b_m according to

$$b_{m-1} = b_0 + x_0\, b_m.$$

For m=4, explicit evaluation of the recursions *via* back substitution shows how each coefficient b_k is naturally computed as a nested polynomial in x_0. This is the most numerically efficient way to evaluate a polynomial because it minimizes the number of mathematical operations.

2.10.2   Horner's Method - Tableau

Horner's Method - Tableau

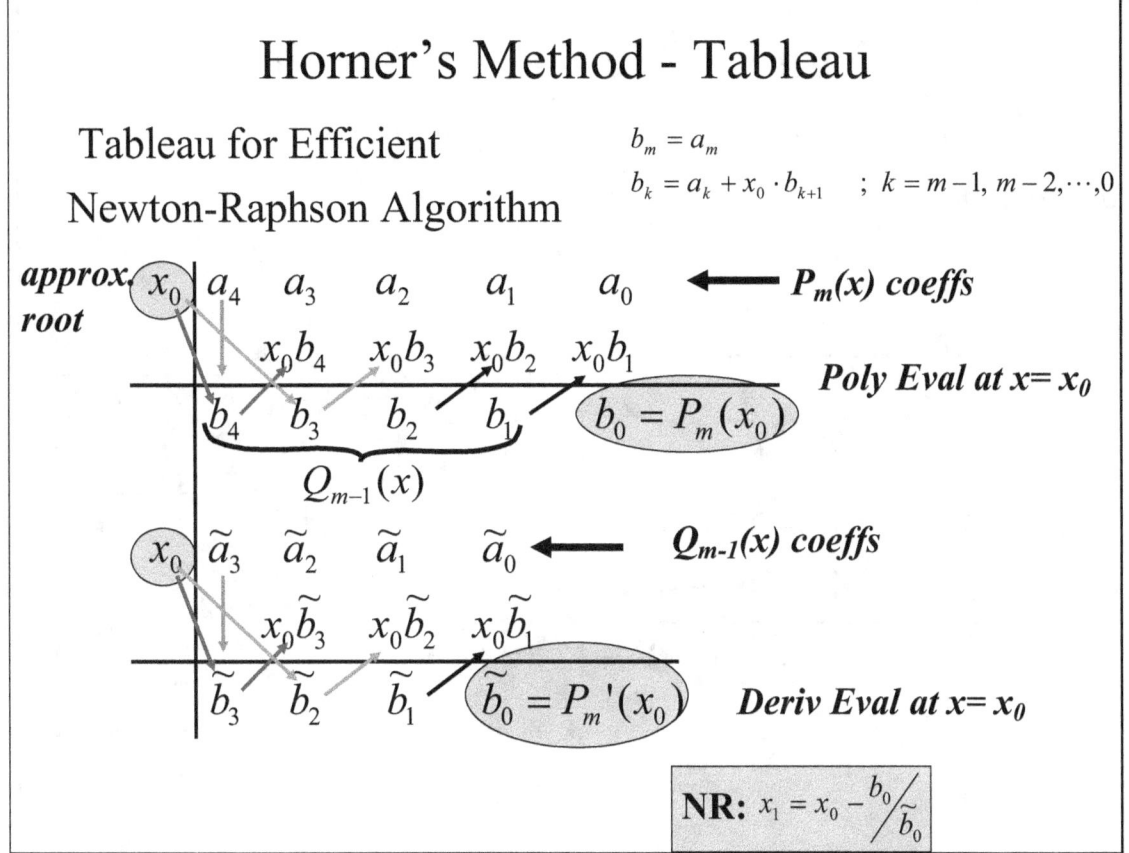

Tableau for Efficient
Newton-Raphson Algorithm

$$b_m = a_m$$
$$b_k = a_k + x_0 \cdot b_{k+1} \quad ; \quad k = m-1, m-2, \cdots, 0$$

The Horner tableau lists the approximate root x_0 on the left and the descending coefficients of the original polynomial $P_m(x)$ from left to right across the top. The arrows in the tableau implement recursions in a very simple manner. The value a_4 is dropped down into position below the line as b_4 ($a_m=b_m$); then x_0 multiplies b_4 and adds to a_3 yielding b_3 ($b_3=x_0*b_4+a_3$) under the line; next x_0 multiplies b_3 and adds to a_2 yielding b_2 ($b_2=x_0*b_3+a_2$) under the line; next x_0 multiplies b_2 and adds to a_1 yielding b_1 ($b_1=x_0*b_2+a_1$) under the line; finally x_0 multiplies b_1 and adds to a_0 yielding b_0 ($b_0=x_0*b_1+a_0$) which is circled and set equal to $P_4(x_0)$ under the line.

The coefficients of $Q_{m-1}(x)$, a cubic in this case, are b_4 b_3 b_2 and b_1. Note b_0 does not belong to the cubic polynomial; it is the remainder of the division and in fact evaluates $P_4(x_0) = b_0$.

We can formally perform a second division by $x-x_0$ by copying the b's and renaming them a-twiddles as shown (this is not necessary as we could operate with the b's directly) We simply perform analogous steps on the new polynomial with 4 a-twiddles to obtain the 3 b-twiddles corresponding to a quadratic and the b_0-twiddle corresponding to the evaluation of the polynomial $Q_{m-1}(x_0)= P_m'(x_0)$ the derivative.

The N-R update iteration is given in the boxed equation as x_0 minus the ratio of the two remainders.

2.10.3   Horner's Method - Example

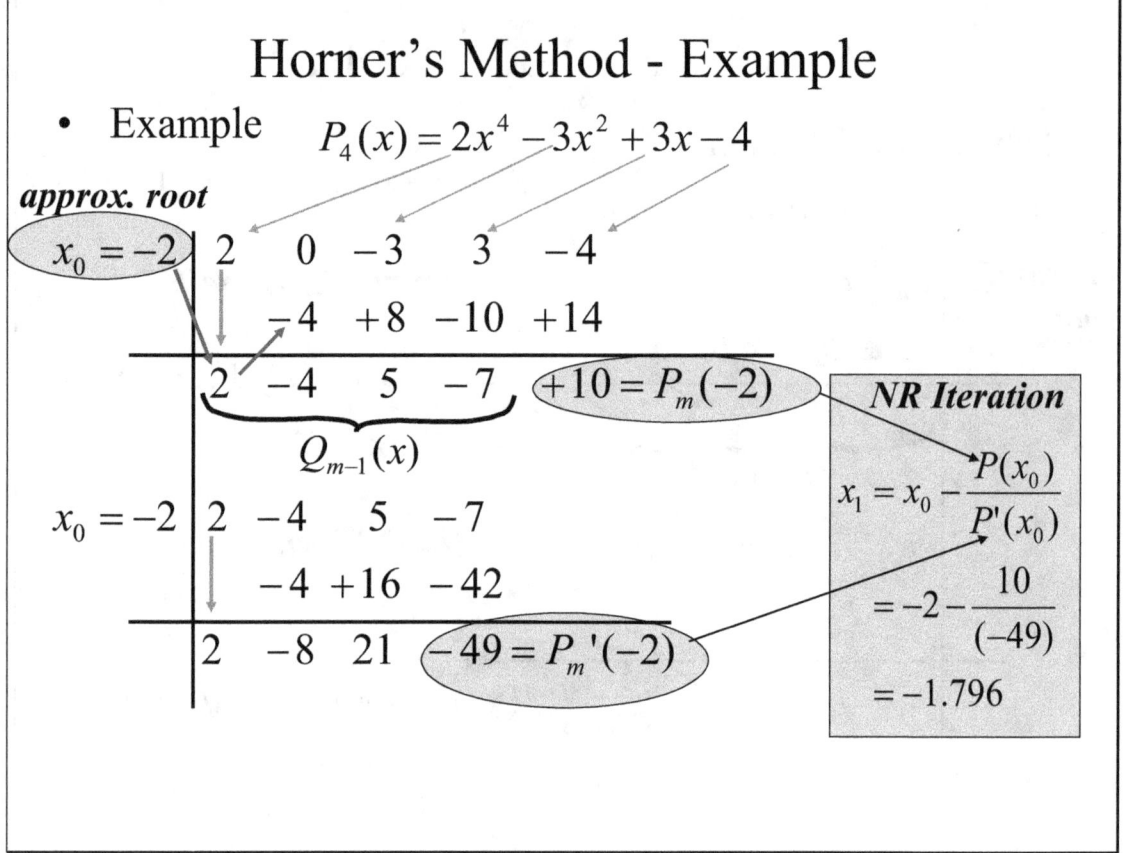

This explicit example follows the Horner Tableau pattern established on the last slide. Note that all degrees of the polynomial must appear, even if the coefficient is zero, *i.e.*, the term "0"x^3 in this example. The approximate root $x_0=-2$ is run through the two synthetic divisions yielding $P_m(-2)$ and $P'_m(-2)$ and a quadratic with 3 coefficients at the bottom of the tableau.

The N-R update yields a new estimate for the root $x_1 = -1.796$; a second pass through the whole tableau using this value - 1.796 will yield two new remainders and a new N-R iterate x_2. This process continues until the N-R correction term $P(x_0) / P'(x_0)$ (*i.e.*, the ratio of polynomial division remainders) becomes less than the desired error, say 10^{-6}.

2.10.4   Computational Efficiency of Horner's Method

Computational Efficiency of Horner's Method

1. Polynom. $P_m(x) = a_0 + a_1 \cdot x^1 + a_2 \cdot \underbrace{x^2}_{x \cdot x} + \cdots + a_m \cdot \underbrace{x^m}_{x \cdot x \cdots x}$

2. Horner recursion relations for b's given a's

$$b_m = a_m$$

$$b_k = a_k + x_0 \cdot b_{k+1} \quad ; \; k = m-1, \, m-2, \cdots, 0$$

Number of Mults (M), Adds(A)

1. Direct Method   2. Horner(Nesting)

$$M = 1+2+3+\cdots+m = \frac{m}{2} \cdot (m+1)$$

$$A = m$$

$$OPs = A+M = \frac{m^2}{2} + \frac{3m}{2} \propto \left(\frac{m^2}{2}\right)$$

$$M = 1 \cdot \sum_{k=0}^{m-1} 1 = m$$

$$A = 1 \cdot \sum_{k=0}^{m-1} 1 = m$$

$$OPs = 2m$$

Example: m = 20 Ops = 230	Ops = 40

Direct evaluation of an mth degree polynomial requires 1 multiplication for the a_1-term, 2 multiplications for a_2-term, ..., m multiplications for a_m-term; thus, the total number of multiplications

$$M = 1+2+3+\ldots+m = m(m+1)/2$$

The sum of m+1 terms requires A=m additions, so that the total number of operations is

$$A+M = m^2/2 + 3m/2 \sim m^2/2$$

Horner Recursion Relations require only 1 multiplication and 1 addition for each index of recursion k=0,..,m-1 yielding a total

$$A+M = 2m$$

Clearly, for large m=1000, the difference is huge Direct$\sim (1000)^2/2 = 5 \times 10^5$ *vs.* Horner $= 2 \times 10^3$

2.10.5   Polynomial Root Finding Methodology

Polynomial Root Finding Methodology

1. ***Horner's Method:*** Evaluate $P(x_0)$ & $P'(x_0)$

2. ***Newton-Raphson Iterations:*** for 1st root until b_0 is "small"

3. ***Deflate Polynomial Degree:*** $\tilde{P}_{m-1}(x) \equiv Q_{m-1}(x)$

4. ***Repeat Steps 1-3:*** on deflated polynomials until estimates of all m roots are found: $\hat{x}_1, \hat{x}_2, \cdots, \hat{x}_m$

5. ***Fine Tuning:*** Use each root found in step #4 as an initial value in a N-R procedure applied to the ***original*** (non-deflated polynomial $P_m(x)$)

Finding all the roots of a polynomial is a problem that arises in many engineering contexts, such as eigenvalue analysis or the numerical solution to differential equations. The computational efficiency of the Horner method together with the fast convergence rate of Newton-Raphson algorithm is a good combination to find roots but requires good start points to find a specific root. The latter fact is not really a drawback if we need to find all the roots, not just a specific one. For well-behaved polynomials the root finding technique outlined here is quite good.

The basic idea in steps #1, #2 is to quickly find a root via Newton-Raphson iterations, stopping when the correction term is "small" and then (step#3) use this approximate root to deflate the polynomial to one of degree m-1. In step#4 we repeat this process (Steps #1-3) until there are approximate estimates of all m roots. The final step#5 performs N-R iterations with the *original polynomial* using the reasonably accurate starting values given by the approximate roots found in step#4.

2.11 *Accelerating Convergence of Iterates*

Accelerating Convergence of Iterates

- Linear Convergent Sequence: $\dfrac{\Delta p_{k-1}}{\Delta p_{k-2}} \cong const. \equiv C_L$

$$p_{k-2}^{new} = p_{k-2} - \frac{(\Delta p_{k-2})^2}{\Delta^2 p_{k-2}} = \{\Delta^2\} p_{k-2}$$

$$\Delta p_{k-2} = p_{k-1} - p_{k-2}$$

$$\Delta^2 p_{k-2} = \Delta(p_{k-1} - p_{k-2}) = (p_k - p_{k-1}) - (p_{k-1} - p_{k-2}) = p_k - 2p_{k-1} + p_{k-2}$$

n	p_n	Δp_n	$\Delta^2 p_n$	$\Delta p_n / \Delta p_{n-1}$	$p_n^{new}=\{\Delta^2\}p_n$	Re-label
4	.51315	.12880 –01	-.62000 –02	*Convergence*	.53991	0
5	.52603	.66800-02		*Test*		1
6	.53271			↓		2

$$p_0^{new} = p_0 - \frac{(\Delta p_0)^2}{\Delta^2 p_0}$$

Once a sequence starts to converge linearly, we can invoke the Aitken's Method defined in the boxed equation to accelerate the rate of convergence. The main issue is to determine when a sequence is converging linearly. The actual error $\varepsilon_k = |p-p_k|$ assumes we know the limit of the sequence "p", which of course we don't. Linear convergence is defined by the ratio of consecutive iterate errors by $\varepsilon_k / \varepsilon_{k-1} = C_L$ a constant. So the question is "how can we establish that the iterates are converging without knowing the answer p"?

The first and second differences between iterates as well as their ratio (shaded column) can be tabulated as we generate iterates in a table. It turns out that this ratio of first differences is a good approximation to the error ratio and this fact allows us establish linear convergence by simply observing when the ratio becomes approximately constant. Once established we start computing a new improved iterate by applying Aitken's formula.

The Aitken improved iterate p_4^{new} requires three terms in the iteration sequence p_n in order to generate two terms in the 1st difference column Δp_n (difference between successive terms in the p_n column) and one term in the 2nd difference column $\Delta^2 p_n$ (difference between the successive terms in the Δp_n column.) The entries in these columns are easily verified; *e.g.*, the second row in the Δp_n column is .53271-.52603=.00668 and the entry in the $\Delta^2 p_n$ .00668-.012880 = -.0062. The Aitken improved iterate is computed (boxed equation) as $p_4^{new} = p_4 - (\Delta p_4)^2 / \Delta^2 p_4 = .51315 - (.012880)^2/(-.0062) = .53991$. Thus in the table we have computed the improved iterate $p_4^{new} = .53991$ from the three iterates p_4, p_5, p_6 in the second column.

2.11.1 Equivalence of $\Delta p_n/\Delta p_{n-1}$ to $\varepsilon_n/\varepsilon_{n-1}$

Equivalence of $\Delta p_n/\Delta p_{n-1}$ to $\varepsilon_n/\varepsilon_{n-1}$

- Linear Convergent Sequence: $\quad \dfrac{\varepsilon_n}{\varepsilon_{n-1}} = const. \equiv C_L$
- Using the definitions for Δp_n, Δp_{n-1} & forming ratio

$$\Delta p_n = p_n - p_{n-1} \qquad \Delta p_{n-1} = p_{n-1} - p_{n-2}$$

$$\frac{\Delta p_n}{\Delta p_{n-1}} = \frac{(p_n - p) - p_{n-1} + p}{(p_{n-1} - p) - p_{n-2} + p} = \frac{\varepsilon_n - \varepsilon_{n-1}}{\varepsilon_{n-1} - \varepsilon_{n-2}} = \frac{\varepsilon_n}{\varepsilon_{n-1}} \cdot \frac{1 - \varepsilon_{n-1}/\varepsilon_n}{1 - \varepsilon_{n-2}/\varepsilon_{n-1}}$$

- Thus

$$\frac{\Delta p_n}{\Delta p_{n-1}} \cong \frac{\varepsilon_n}{\varepsilon_{n-1}} \cdot \underbrace{\frac{1 - 1/C_L}{1 - 1/C_L}}_{=1} = \frac{\varepsilon_n}{\varepsilon_{n-1}}$$

This slide uses the definition of linear convergence to establish the approximate equality between the error ratio $\varepsilon_n/\varepsilon_{n-1}$ and the ratio between the differences Δp_n of successive iterates, *viz.*, $\Delta p_n/\Delta p_{n-1}$. We rewrite the ratio of iterate differences $p_n - p_{n-1}$ by adding zero (in the form of the identity $p - p = 0$) to both numerator and denominator; the resulting expression may be rearranged into a from which now involves ratios of the true but unknown errors $\varepsilon_{n-1}/\varepsilon_n$ and $\varepsilon_{n-2}/\varepsilon_{n-1}$ as shown in the slide. Since the sequence is converging linearly, both of these ratios may be approximated by the same constant C_L. This, in turn, allows us to cancel the numerator and denominator of the 2nd term, and leaves the term $\varepsilon_n/\varepsilon_{n-1}$ on the rhs, thus establishing the approximate relation between it and $\Delta p_n/\Delta p_{n-1}$.

2.11.2   Aitken's Method: g(x)=ln(2cosx)

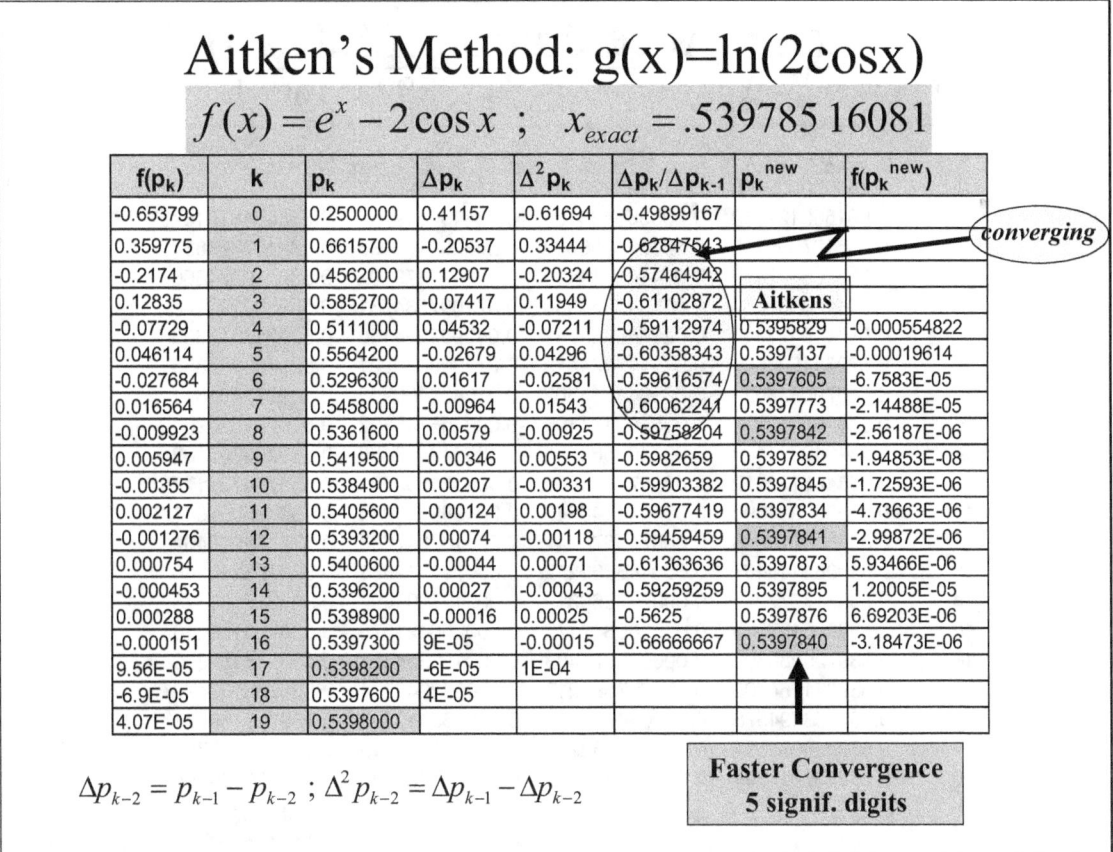

Aitken's Method: g(x)=ln(2cosx)

$$f(x) = e^x - 2\cos x \;\; ; \;\; x_{exact} = .539785\,16081$$

f(p_k)	k	p_k	Δp_k	Δ²p_k	Δp_k/Δp_{k-1}	p_k^new	f(p_k^new)
-0.653799	0	0.2500000	0.41157	-0.61694	-0.49899167		
0.359775	1	0.6615700	-0.20537	0.33444	-0.62847543		
-0.2174	2	0.4562000	0.12907	-0.20324	0.57464942		
0.12835	3	0.5852700	-0.07417	0.11949	-0.61102872	**Aitkens**	
-0.07729	4	0.5111000	0.04532	-0.07211	-0.59112974	0.5395829	-0.000554822
0.046114	5	0.5564200	-0.02679	0.04296	-0.60358343	0.5397137	-0.00019614
-0.027684	6	0.5296300	0.01617	-0.02581	-0.59616574	0.5397605	-6.7583E-05
0.016564	7	0.5458000	-0.00964	0.01543	-0.60062241	0.5397773	-2.14488E-05
-0.009923	8	0.5361600	0.00579	-0.00925	-0.59758204	0.5397842	-2.56187E-06
0.005947	9	0.5419500	-0.00346	0.00553	-0.5982659	0.5397852	-1.94853E-08
-0.00355	10	0.5384900	0.00207	-0.00331	-0.59903382	0.5397845	-1.72593E-06
0.002127	11	0.5405600	-0.00124	0.00198	-0.59677419	0.5397834	-4.73663E-06
-0.001276	12	0.5393200	0.00074	-0.00118	-0.59459459	0.5397841	-2.99872E-06
0.000754	13	0.5400600	-0.00044	0.00071	-0.61363636	0.5397873	5.93466E-06
-0.000453	14	0.5396200	0.00027	-0.00043	-0.59259259	0.5397895	1.20005E-05
0.000288	15	0.5398900	-0.00016	0.00025	-0.5625	0.5397876	6.69203E-06
-0.000151	16	0.5397300	9E-05	-0.00015	-0.66666667	0.5397840	-3.18473E-06
9.56E-05	17	0.5398200	-6E-05	1E-04			
-6.9E-05	18	0.5397600	4E-05				
4.07E-05	19	0.5398000					

converging

$$\Delta p_{k-2} = p_{k-1} - p_{k-2} \;\; ; \;\; \Delta^2 p_{k-2} = \Delta p_{k-1} - \Delta p_{k-2}$$

**Faster Convergence
5 signif. digits**

Here is an explicit example which finds a root of $f(x) = e^x - \ln(2\cos(x))$ using the fixed point method with function $g(x) = \ln(2\cos(x))$ and an initial guess $p_0 = .2500000$. The table shows the sequence p_n, its 1st and 2nd differences, their ratio and the Aitken's p_n^{new}; also the columns on either end show the functional evaluations $f(p_n)$ and $f(p_n^{new})$ for the two sequences.

The yellow oval shows that the ratio column is converging to a "constant" and the decision is made to start computing the Aitken's sequence starting at k=4. Comparing these two sequences shows that the Aitken's sequence actually does converge faster reaching 5 sd in the k=16th iteration. The functional evaluations in the 1st and last columns also verify that $f(x)$ reaches 10^{-6} for the Aitken method, while it remains at 10^{-5} for the FP method.

2.11.3 Aitken's Acceleration (MatLab)

Aitken's Acceleration (MatLab)

$$f(x) = e^x - 2\cos x \; ; \quad x_{exact} = .53978516081$$

k	p_k	Δp_k	$\Delta^2 p_k$	p_k^{new}
0	0.250000000000	0.411566129312	-0.616934692507	0.524561766188
1	0.661566129312	-0.205368563195	0.334436819547	0.535454875157
2	0.456197566118	0.129068256352	-0.203230170465	0.538166767637
3	0.585265822470	-0.074161914113	0.119474479102	0.539230974306
4	0.511103908357	0.045312564990	-0.072097534731	0.539582393143
5	0.556416473346	-0.026784969741	0.042952294160	0.539713416663
6	0.529631503605	0.016167324419	-0.025808638221	0.539759212655
7	0.545798828024	-0.009641313802	0.015433490998	0.539775891811
8	0.536157514222	0.005792177196	-0.009256745802	0.539781824254
9	0.541949691418	-0.003464568606	0.005542360486	0.539783965162
10	**0.538485122812**	**0.002077791880**	**-0.003321938525**	**0.539784731177**
11	0.540562914692	-0.001244146645	0.001989825086	0.539785006681
12	0.539318768046	0.000745678441	-0.001192347682	0.539785105462
13	0.540064446487	-0.000446669242	0.000714319561	0.539785140946
14	0.539617777245	0.000267650319	-0.000427997492	0.539785153678
15	0.539885427565	-0.000160347172	0.000256421575	0.539785158250
16	0.539725080392	0.000096074403	-0.000153634621	0.539785159890
17	0.539821154795	-0.000057560217	0.000092047274	0.539785160479
18	0.539763594578	0.000034487056	-0.000055149347	0.539785160691
19	0.539798081634	-0.000020662291	0.000033041919	0.539785160767
20	**0.539777419344**	**0.000012379628**	**-0.000019796703**	**0.539785160794**
	5 sd			10 sd

If we "know" or "take the chance" that the sequence is convergent, we can start computing Aitken's p_n^{new} as soon as we have three values. Thus the 1st term in the last column starts with p_0^{new} which is computed from the terms p_0, p_1, and p_2. (See MatLab script on Slide#5-5)

Comparing the two sequences at the k=20 shows that the FP method only achieves 5 sd of accuracy, while the Aitken "gambit without testing" yields 10 sd. Clearly, taking a chance and starting the Aitken sequence immediately has an upside producing 5 more sd of accuracy!

2.11.4 Aitken's Method – Math Details

Aitken's Method – Math Details

Linear convergence:

$$\frac{\Delta p_{k-1}}{\Delta p_{k-2}} \cong const. \equiv C_L$$

Each ratio is bounded by C_L so we may write the following sequence and subs into →

$$\Delta p_{k-1} = C_L \cdot \Delta p_{k-2}$$

$$\Delta p_k = C_L \cdot \Delta p_{k-1} = C_L \cdot (C_L \cdot \Delta p_{k-2}) = C_L^2 \cdot \Delta p_{k-2}$$

$$\Delta p_{k+1} = C_L \cdot \Delta p_k = C_L \cdot (C_L^2 \cdot \Delta p_{k-2}) = = C_L^3 \cdot \Delta p_{k-2}$$

$$\vdots$$

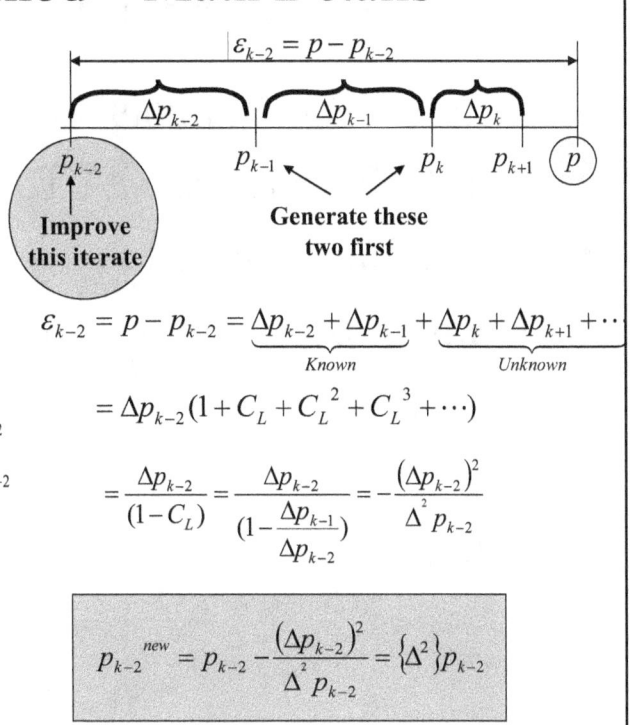

$$|\varepsilon_{k-2} = p - p_{k-2}$$

Improve this iterate

Generate these two first

$$\varepsilon_{k-2} = p - p_{k-2} = \underbrace{\Delta p_{k-2} + \Delta p_{k-1}}_{Known} + \underbrace{\Delta p_k + \Delta p_{k+1}}_{Unknown} + \cdots$$

$$= \Delta p_{k-2}(1 + C_L + C_L^2 + C_L^3 + \cdots)$$

$$= \frac{\Delta p_{k-2}}{(1 - C_L)} = \frac{\Delta p_{k-2}}{(1 - \frac{\Delta p_{k-1}}{\Delta p_{k-2}})} = -\frac{(\Delta p_{k-2})^2}{\Delta^2 p_{k-2}}$$

New Aitkens' Sequence convergences more rapidly

$$p_{k-2}^{new} = p_{k-2} - \frac{(\Delta p_{k-2})^2}{\Delta^2 p_{k-2}} = \{\Delta^2\} p_{k-2}$$

Here is an intuitive proof of the Aitken's formula. The figure shows that the difference between the convergent value "p" and the iterate p_{k-2} can be broken up into an infinite sum of "Δp_k"s , only two of which are actually known. However, if the sequence converges, then all of these terms can be written in terms of the constant ratio of iterate differences, and as shown in the equations under the linear convergence box on the upper left, all the unknown "Δp_k"s may be expressed in terms of the linear convergence ratio C_L.

Thus, the infinite sum on RHS of the slide becomes a geometric series in C_L whose sum can be written down immediately. Now substituting the value of C_L obtained from the ratio of the two known "Δp_k"s we obtain Aitken's formula for p_{k-2}^{new} expressed in terms of prior known iterates.

3 Polynomial Interpolation

Polynomial Interpolation

- Lagrange Polynomials
- Divided Difference & Δ Tables
- Inverse Interpolation
- Hermite Polynomials, Cubic Splines

Here we discuss methods of taking a limited number of tabulated values of a function, interpolating between them, and extrapolating beyond the first and last data points. The basic method is to fit various degree polynomials through small groups of the tabulated points in the smoothest manner possible. It may seem there is just one way to interpolate or extrapolate, inasmuch as n+1 values from a tabular function uniquely determines a polynomial of degree n. However, the numerical efficiencies, orthogonality properties, degree of oscillation, smoothness, and other specific properties associated with different polynomials makes the choice of a given polynomial anything but trivial.

3.1 *Fitting a Curve to Tabulated Values*

Fitting a Curve to Tabulated Values

Given a set of n nodal values (x_k , $f(x_k)$) of an unknown function $f(x)$
Estimate $f(x)$ between (interpolate) and beyond (extrapolate) nodes

1. The interpolating polynomial $P_n(x)$ is *uniquely* determined by the set of n nodal values

2. The unknown function $f(x)$ is not!

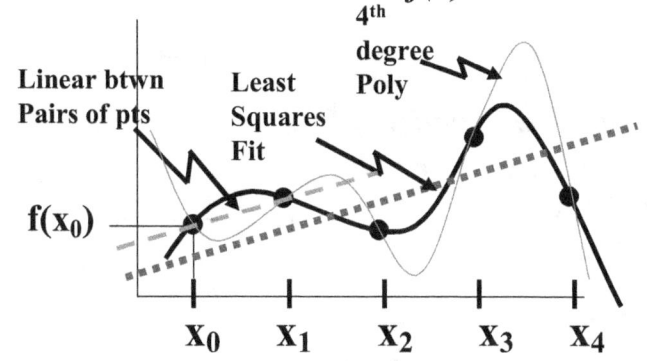

Bessel Fcn 1ˢᵗ kind		
k	x_k	$f(x_k)$
0	1.0	.7651977
1	1.3	.6200860
2	1.6	.4554022
3	1.9	.2818186
4	2.2	.1103623

One key fact to understand is that, although the interpolating polynomial is uniquely determined by the set of n nodal values (illustrated as black dots in the figure), the unknown function f(x) is only required to go through the nodes and is free to oscillate in any manner in between. For example, the red and black curves are quite different even through they go through the same nodal points.

If for some physical reason we expect a straight line it would be foolish to fit it with a polynomial; instead we would use a least squares fit to the slope and intercept parameters of a straight line. Alternately, we may just be seeking a polynomial approximation that generates a guaranteed fixed accuracy between tabulated points such as those displayed in the Bessel function table.

3.2 *Polynomial Properties*

Polynomial Properties

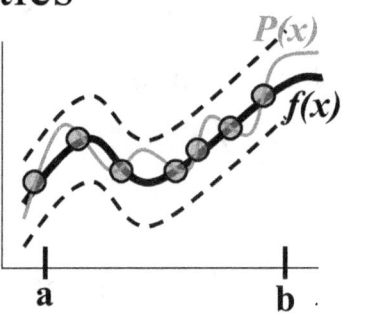

1. **Weierstrass Approximation Theorem:**
   "polynomials uniformly approximate continuous functions" $\{1, x^1, x^2, x^3, \cdots\}$

2. **Easy to work with**
   multiply, add, divide, differentiate, and integrate

3. **Truncation error is** $\quad \varepsilon_{trunc.} \equiv f(x) - P_n(x)$
   - **Made explicit**
   - **Easily analyzed**
   $$= f^{(n+1)}(\xi)\frac{(x-x_0)^{n+1}}{(n+1)!}; \qquad \xi(x) \in [x_0, x]$$

4. **Orthonormal (orthogonal & normalized)**
   - **Linear Vector Space ideas allow "clean"**
     - **Separation of orders of magnitude**
     - **Computation of coefficients** $\left\{1, x, \dfrac{3x^2-1}{2}, \cdots\right\}$
     - **Error analysis**
   - **Improper integrals and weighting functions**
   - **Adapt to specific coordinate system symmetries**
   - **Hermite, Laguerre, Tschebyshev, Legendre, ...**

The main reason for using polynomials is that the Weierstrass Approximation Theorem states that the infinite set of polynomials $\{1, x, x^2, ... \}$ uniformly approximates continuous functions over any finite interval [a,b] . Other obvious reasons for their use is that they are easy to manipulate algebraically, and they are easy to differentiate and integrate. Theorems analogous to Taylor's theorem allow us to bound the truncation error. Special types of orthogonal and normalized (orthonormal) polynomials allow us to apply linear vector space ideas and facilitate error analysis because orthogonality provides clean separation of the truncation errors by orders of magnitude. Many of these functions are adapted to specific coordinate systems and/or differential equations and thus have properties that make them uniquely suited to special classes of problems.

Finally, the orthogonality relations usually involve exponential, Gaussian, or other weighting functions that also appear in "improper integrals." Such integrals with infinities in their limits or in their integrands are difficult to evaluate accurately without resorting to special adaptive methods that use the orthogonality integrals associated with these special polynomials.

3.2.1 Why Different Polynomials?

Why Different Polynomials?

1. Usage
   - Generate Tables
   - Approximate a function over an interval
   - Find a root
   - Derivatives, Integrals
   - Smooth appearance of curves "duck"
   - Computer Graphics (drawing curves)

2. Robustness
   - Near singularities
   - Convergence circle - Over relatively large regions
   - Computational efficiency
     - Truncation - Number of terms for given error tolerance
     - Number of Ops – Round off error
   - Recursion relations- generate polynomials on-the-fly
   - Orthonormal – separate error orders-of-magnitude
   - Ease of adding new data
   - Ease of using auxilliary data (position and velocity)

Instead of using the canonical Weierstrass polynomial basis functions $\{1, x, x^2, ... \}$ we may choose linear combinations of this canonical set which are associated with specific differential equations and hence physical phenomena governed by them. These new sets of basis polynomials (with names like Legendre, Laguerre, Hermite, Tschebyshev) are orthonormal relative to a specific weighting function associated with the differential equation and hence are the natural set for the specific physical phenomena.

We have already mentioned the use of orthonormal polynomials to evaluate improper integrals. There are yet other reasons to choose different polynomials based on their intended usage and desired robustness.

To be more specific, there are special polynomials that (i) produce a uniform error across a range of values in a table of interpolated values, (ii) are specially adapted to the solution of a specific class of problems, and (iii) that can create very smooth curves (drawing) as well as to change shape in response to mouse manipulation (as in computer graphics).

Robustness near singularities, large regions of convergence, computational efficiency, ease of adding new data nodes without the need for complete recalculation are also important. The ability to go beyond a simple functional table to include both position and velocity is also accomplished with a special choice of polynomial. The set of tables which follow give an overview of these polynomials and some of the properties that make them special.

3.2.2 Polynomial Overview - 1

Polynomial Overview-1

- **Vandermonde (Muller)**
  - Multiple points
  - No derivatives
  - Matrix formulation - nearly singular matrix
  - **Problem:** No Error Estimate

$$P_n^V(x) = a_n x^n + a_{n-1} x^{n-1} + \cdots + a_1 x + a_0$$

$$\begin{bmatrix} f_0 \\ f_1 \\ \vdots \\ f_n \end{bmatrix} = \begin{bmatrix} x_0^n & \cdots & x_0 & 1 \\ x_1^n & \cdots & x_1 & 1 \\ \vdots & & \vdots & \vdots \\ x_n^n & \cdots & x_n & 1 \end{bmatrix} \cdot \begin{bmatrix} a_n \\ \vdots \\ a_1 \\ a_0 \end{bmatrix}$$

Vandermonde Matrix

- **Taylor**
  - Single point
  - Multiple derivatives
  - Series Expansion
  - **Problem:** Good at nearby points only - Local

$$f(x) \approx P_n^T(x) \equiv \sum_{k=0}^{n} f^{(k)}(x_0) \frac{(x - x_0)^k}{k!}$$

- **Lagrange**
  - Multiple points
  - No derivatives
  - Algorithmic induction –work directly with numbers
  - **Problem**: Unwanted oscillations – over fitting

$$f(x) \approx P_n^L(x) \equiv \sum_{\alpha=0}^{n} L_{n,\alpha}(x) \cdot f(x_\alpha)$$

$$L_{n,\alpha}(x) \equiv \frac{(x - x_0)}{(x_\alpha - x_0)} \cdot \frac{(x - x_1)}{(x_\alpha - x_0)} \cdots \frac{\overset{\alpha}{1}}{1} \cdots \frac{(x_n - x_0)}{(x_\alpha - x_0)}$$

- **Hermite**
  - Multiple points
  - One derivative Takes advantage of measured derivatives (velocities)
  - Series Expansion
  - **Problem**: More OPs – RO error

$$f(x) \approx H_{2n+1}(x) \equiv \sum_{k=0}^{n} \left(H_{n,k}(x) \cdot \underbrace{f(x_k)}_{\text{nodal vals.}} + \hat{H}_{n,k}(x) \cdot \underbrace{f'(x_k)}_{\text{nodal derivs.}} \right)$$

$$H_{n,k}(x) \equiv \left[1 - 2(x - x_k) L'_{n,k}(x_k)\right] \cdot L^2_{n,k}(x)$$

$$\hat{H}_{n,k}(x) \equiv (x - x_k) \cdot L^2_{n,k}(x)$$

This slide shows a progression of polynomials characterized by the number nodal points, the number of functional evaluations and/or derivatives that are needed in their construction.

Vandermonde is a generalization of the Muller method which fits multiple nodal points by solving a nearly singular matrix equation. It requires no derivatives and is unique in that it makes no error estimate

Taylor *fits a single functional value and the first n derivatives* at a single node to construct a polynomial of degree n. It provides a theoretical estimate of the truncation error and yields good "local" estimates, but its accuracy rapidly deteriorates away from the single nodal value.

Lagrange *fits the functional values (but no derivatives)* at n+1 points to construct a polynomial of degree n. It provides a theoretical estimate of the truncation error and yields good estimates in the extended region covered by the nodes; however, as the number of nodes (degree) increases, the polynomial oscillates because it must pass through all of its nodal points.

Hermite *fits the functional values as well as first derivative* at n+1 points to construct a polynomial of degree 2n+1. It provides a theoretical estimate of the truncation error and yields excellent estimates in the extended region covered by the nodes; however, it is computationally expensive and subject to large round-off errors.

3.2.3 Vandermonde, Taylor, Lagrange, Hermite

Vandermonde, Taylor, Lagrange, Hermite

Name	*Polynomial*	*Truncation Error*
Vander-monde	$P_n^V(x) = a_n x^n + a_{n-1} x^{n-1} + \cdots + a_1 x + a_0$	No Trunc. Error Estimate
Taylor	$P_n^T(x) \equiv \sum_{k=0}^{n} f^{(k)}(x_0) \dfrac{(x-x_0)^k}{k!}$	$R_n^T = \dfrac{f^{(n+1)}(\xi)}{(n+1)!} \cdot (x-x_0)^{n+1}$
Lagrange	$P_n^L(x) \equiv \sum_{k=0}^{n} L_{n,k}(x) \cdot f(x_k)$	$R_n^L(x,\xi) = \dfrac{(x-x_0) \cdot (x-x_1) \cdots (x-x_n)}{(n+1)!} \cdot f^{(n+1)}(\xi)$
Hermite	$H_{2n+1}(x) \equiv \sum_{k=0}^{n} \left(H_{n,k}(x) \cdot \underbrace{f(x_k)}_{\text{nodal vals.}} + \hat{H}_{n,k}(x) \cdot \underbrace{f'(x_k)}_{\text{nodal derivs.}} \right)$	$R_n^H(x,\xi) = \dfrac{(x-x_0)^2 \cdot (x-x_1)^2 \cdots (x-x_n)^2}{(2n+2)!} \cdot f^{(2n+2)}(\xi)$

The explicit analytic form and truncation error are given for each polynomial.

Vandermonde is a simple polynomial with coefficients determined by the solution of a matrix equation and has no error estimate.

Taylor polynomial starts with the functional value at the single point $f(x_0)$ (k=0 term) and then increments this value by the product terms $f^{(k)}(x_0)$ $(x-x_0)^k$ /k! summed from k=1,2, ..., n. The truncation error is equal to the "next term" in the series with the $(n+1)^{th}$ derivative evaluated at an unknown point in the interval of convergence $f^{(n+1)}(\xi)$.

Lagrange polynomial of degree n is formed by the sum of products of the functional values at each node $\{f(x_0), f(x_1), ..., f(x_n)\}$ with their associated n^{th} degree "basis" polynomial $\{L_{n,0}(x), L_{n,1}(x), ... , L_{n,n}(x)\}$ as shown by the formula. These n+1 basis polynomials $L_{n,k}(x)$ all evaluate to unity at their associated nodal point and evaluate to zero at any other nodal points; thus for example $L_{n,0}(x=x_0) =1$ at its nodal point x_0 and at other points $L_{n,0}(x=x_1) =0$, $L_{n,0}(x=x_2) =0$, ... , $L_{n,0}(x=x_n)=0$; the same is true for the other $L_{n,k}(x)$, *viz.*, $L_{n,1}(x=x_1) =1$, ..., $L_{n,n}(x=x_n)=1$, and otherwise zero. Thus the functional values $f(x_k)$ are weighting factors that boost the Lagrange polynomial nodal value from 1 to its correct numerical value at the kth nodal point. The truncation error is similar to that for Taylor as it uses the $(n+1)^{th}$ derivative evaluated at an unknown point; however, instead of multiplying the derivative term by $(x-x_0)^{n+1}$ /(n+1)! it distributes this among all points as $(x-x_0)$ $(x-x_1)... (x-x_n)$ /(n+1)!.

Hermite polynomial of degree 2n+1 is a rather complex sum of pairs of terms that multiply the functional value $f(x_k)$, and its derivative $f'(x_k)$, by two different polynomials. Each term simultaneously insures that the function and its derivative go through their nodal values. The truncation error is hybrid Taylor/Lagrange since it evaluates the $(2n+2)^{th}$ derivative $f^{(2n+2)}(\xi)$ at ξ and also follows Lagrange by distributing the error among all points as $(x-x_0)^2$ $(x-x_1)^2... (x-x_n)^2$ /(2n+2)!.

3.2.4   Polynomial Overview - 2

Polynomial Overview-2

• **_Osculating Polynomials_** Generalization of Hermite
  – Multiple points
  – Multiple derivatives (n=1 is Hermite)
  – Series Expansion
  – **_Problem_**: More elaborate computations – RO error

• **_Tschebychev_**
  – Orthornormal
  – Uniformly distributed over[-1,1]
  – Good coverage ➜
  – Economization ("almost" Taylor Series)
  – **_Problem_**: Need to map to interval [-1,1]

• **_Cubic Splines_**
  – Break up into subsplines every pair of points
  – Match f(x) , 1st and 2nd Derivatives everywhere.
  – Very smooth curves "Duck"
  – **_Problem_**: Elaborate computation – RO error ➜Nesting

• **_Bezier Curves_**
  – Parametric curve representation
  – Hermite representation for each coordinate
  – Guide Points - Used in computer drawing

This slide continues the progression beyond Hermite to special use polynomials.

Osculating polynomials extend the Hermite polynomial concept in a straight forward manner by including the values of 2nd and higher order derivatives at the n+1 nodes. Osculating polynomials will give estimates that conform more closely to the underlying function in an extended region, but are computationally expensive and subject to large round-off errors by virtue of the increased number of operations needed to evaluate the polynomial. We will not consider them further.

Tschebyshev has nodal points located at *the n zeros of the nth degree Tschebyshev orthonormal basis polynomials*. The method provides a theoretical estimate of the truncation error and yields estimates that converge uniformly over the entire interval. However, a coordinate transformation that maps the interval [a,b] to [-1,1] must be made prior to constructing the polynomial. It also has increased computational efficiency relative to a Taylor polynomial *via* " Tschebyshev Economization."

Cubic Spline fits a table of nodal values with contiguous cubic polynomials that match functional values and 1st and 2nd derivatives across the sub-spline boundaries yielding very smooth curves and accurate shape representation, but requires elaborate computation and is subject to large round-off errors.

Bezier Curve is a parametric representation of a curve [x(t), y(t)] is obtained by computing a Hermite polynomial from two nodal tables, one for x(t) and the other for y(t). This Bezier curve representation has two free parameters which allow us to arbitrarily choose two "guide points" that change the curve's shape.

3.2.5 Hermite, Laguerre, Tschebyshev, Legendre

Hermite, Laguerre, Tschebyshev, Legendre

Name	Normalization	Recursion
Hermite	$\int_{-\infty}^{\infty} e^{-x^2} \dfrac{H_m(x) \cdot H_n(x)}{\sqrt{2^m m! \sqrt{\pi}} \cdot \sqrt{2^n n! \sqrt{\pi}}} dx = \delta_{mn}$	$H_{k+1} = 2xH_k - 2kH_{k-1}$ $H_0 = 1 ; H_1 = 2x ; H_2 = (4x^2 - 2)$
Laguerre	$\int_0^{\infty} e^{-x} L_m(x) \cdot L_n(x) dx = \delta_{mn}$	$L_{k+1} = (-x + 2k + 1)L_k - kL_{k-1}$ $L_0 = 1; L_1 = x; L_2 = (x^2 - 4x + 2)/2$
Tscheby	$\int_{-1}^{+1} \dfrac{T_m(x) \cdot T_n(x)}{\sqrt{1-x^2}} dx = \begin{cases} 0 & m \neq n \\ \pi/2 & m = n \neq 0 \\ \pi & m = n = 0 \end{cases}$	$T_k = 2xT_k - T_{k-1}$ $T_0 = 1 ; T_1 = x ; T_2 = (2x^2 - 1)$
Legendre	$\int_{-1}^{+1} P_m(x) \cdot P_n(x) dx = \delta_{mn}$	$(k+1)P_{k+1} = (2k+1)xP_k - kP_{k-1}$ $P_0 = 1 ; P_1 = x ; P_2 = (3x^2 - 1)/2$

Here is a table of well known orthogonal polynomials, displaying their normalization / orthogonality integrals and their recursion relations. The integrals are characterized by their respective weighting functions

$$w(x) = \{e^{-x^2}, e^{-x}, (1-x^2)^{-1/2}, 1\},$$

and associated orthogonality intervals

$$[a,b] = \{(-\infty, \infty), [0, \infty), [-1, +1], [-1, +1]\}.$$

The symbol δ_{mn} is called the kronecker delta and is defined to be 1 when the two indices are equal and 0 otherwise. Thus the integral serves as normalization of the squared polynomial to unity when m = n and as the orthogonality relation when the two are different m $\neq$ n.

The two term recursion relations all yield the third when the first two are known. For example, the first two Tschebyshev polynomials are $T_0 = 1$ and $T_1 = x$; the recursion relation for k=1,2,3 yields

$$T_2 = 2xT_1 - T_0 = 2x(x) - 1 = 2x^2 - 1;$$
$$T_3 = 2xT_2 - T_1 = 2x(2x^2 - 1) - x = 4x^3 - 3x ;$$
$$T_4 = 2xT_3 - T_2 = 2x(4x^3 - 3x) - (2x^2 - 1) = 8x^4 - 8x^2 + 1;$$

We will see that each of these polynomials have unique applications for numerical computation (See Slides#3-31, 4-26 to 4-28).

3.3 Taylor Polynomials

Taylor Polynomials

- n^{th} degree polynomial $\quad P_n^T(x) \equiv \sum_{k=0}^{n} f^{(k)}(x_0) \dfrac{(x-x_0)^k}{k!}$

-

- Trunc. Error $\quad R_n^T = \dfrac{f^{(n+1)}(\xi)}{(n+1)!} \cdot (x-x_0)^{n+1}$

- Example: $f(x) = (1+x)^{1/2}$ about $x_0 = 0$

$$P_3^T(x) = 1 + \frac{1}{2}(x-0) - \frac{1}{8}(x-0)^2 + \frac{1}{16}(x-0)^3$$

$$P_3^T(0.1) = 1 + \underbrace{\underbrace{\frac{1}{2}(0.1)}_{P_1=1.05} - \frac{1}{8}(0.1)^2 + \frac{1}{16}(0.1)^3}_{P_2=1.0488125}$$

$$= 1.0488125$$

$$\varepsilon = \left| f(0.1) - P_3^T(0.1) \right| = \boxed{3.65 \times 10^{-6}}$$

$$\left| R_3(0.1) \right| = \left| f^{(4)}(\xi) \frac{(0.1)^4}{(4)!} \right| \qquad \xi \in [0, 0.1]$$

$$= \left| \frac{-15}{16}(1+\xi)^{-7/2} \frac{(0.1)^4}{(4)!} \right|$$

$$\leq 3.91 \times 10^{-6}$$

compare

Here is a basic example of Taylor's Theorem. The 3rd degree Taylor Polynomial about $x_0=0$ is computed and then evaluated at a nearby point x = 0.1. The actual error is obtained by subtracting the evaluated polynomial $P_3(.1) = 1.0488125$ from the "exact" value of $f(.1) = (1+.1)^{1/2} = 1.0488088$ which yields 3.65×10^{-6}. The Taylor theorem states that the truncation error for the polynomial approximation to the function $P_3(x)$ is equal to the next term in the series with the $(3+1)^{st}$ derivative evaluated at an unknown point in the interval of interest. In this case, we wish to maximize $f^{(iv)}(\xi)$ in the interval $\xi \in [0, 0.1]$, which clearly occurs at $\xi=0$ and yields the upper bound 3.91×10^{-6}; this "upper bound" is greater than the actual error as it should be. A more natural interval might be to specify a bound for all $x \in [-0.1, 0.1]$, which yields a worst case at $\xi=-0.1$ and an upper bound 5.65×10^{-6}.

3.3.1 Taylor Polynomials - Example

Taylor Polynomials - Example

- **Example:** $f(x) = \dfrac{1}{x}$

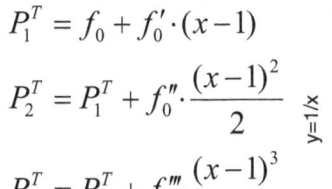

$$P_1^T = f_0 + f_0' \cdot (x-1)$$

$$P_2^T = P_1^T + f_0'' \cdot \frac{(x-1)^2}{2}$$

$$P_3^T = P_2^T + f_0''' \cdot \frac{(x-1)^3}{6}$$

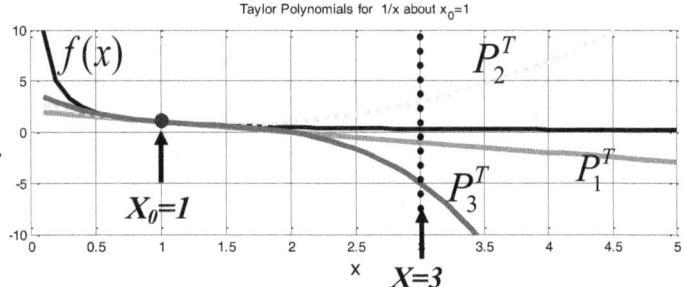

- **Accuracy**
  - Increases with degree near $x_0=1$
  - Does not improve with degree near $x=3$
  - Useful only near a *single point*

The function f(x) = 1/x is plotted as the solid black curve and Taylor polynomials of degree 1, 2, and 3 about x_0 = 1 are computed and plotted as red, green, and blue curves respectively. This plot was done using MatLab and the program appears at the end of these slides. (See MatLab script on Slide#5-6)

It is obvious from the plots that increasing the degree of the Taylor polynomial yields better approximations to the function f(x) for evaluation points near $x_0=1$. However, if we evaluate these three Taylor polynomials at points far from x_0, say along the vertical black dotted line at x = 3, there is no way to predict which one is the best approximation. A glance at the figure shows that for this particular f(x) = 1/x, the linear (red) polynomial P_1(x=3) turns out to be the closest to the actual value f(x = 3).

We conclude that higher degree Taylor polynomials always decrease the truncation error and therefore give a higher accuracy result provided we evaluate the polynomial at points near the point x_0, *i.e.,* we must have (x- x_0) $\ll$ 1; otherwise, accuracy of different degree expansions is unpredictable as illustrated in the figure. (See Matlab script for the plots on Slide#5-6.)

3.4 *Lagrange Interpolating Polynomials*

Lagrange Interpolating Polynomials

- Linear Interpolation

$$P_1^L(x) = f(x_0) \cdot a(x) + f(x_1) \cdot b(x)$$

$$a(x) = a_0 + a_1 x$$
$$b(x) = b_0 + b_1 x$$

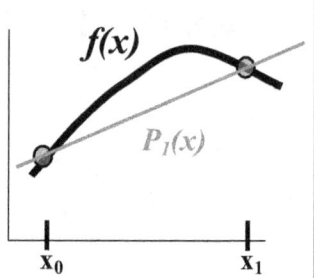

$f(x)$

$P_1(x)$

Require: $\left. \begin{array}{l} a(x_0) = 1 = a_0 + a_1 x_0 \\ a(x_1) = 0 = a_0 + a_1 x_1 \end{array} \right\} \Rightarrow a_0 = \dfrac{-x_1}{x_0 - x_1} \; ; \; a_1 = \dfrac{1}{x_0 - x_1}$

$$\Rightarrow a(x) = \left(\frac{-x_1}{x_0 - x_1} \right) + \left(\frac{1}{x_0 - x_1} \right) \cdot x = \frac{x - x_1}{x_0 - x_1} \; ; \text{ similarly } b(x) = \frac{x - x_0}{x_1 - x_0}$$

$x_0 \qquad x_1$

$$\therefore \; P_1(x) = f(x_0) \cdot \underbrace{\frac{(x - x_1)}{(x_0 - x_1)}}_{\equiv L_{1,0}(x)} + f(x_1) \cdot \underbrace{\frac{(x - x_0)}{(x_1 - x_0)}}_{\equiv L_{1,1}(x)}$$

- Quadratic Interpolation

$$P_2^L(x) = f(x_0) \cdot a(x) + f(x_1) \cdot b(x) + f(x_2) \cdot c(x)$$

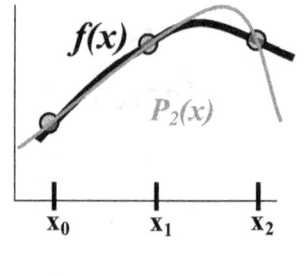

$f(x)$

$P_2(x)$

$$P_2(x) = f(x_0) \cdot \underbrace{\frac{(x - x_1) \cdot (x - x_2)}{(x_0 - x_1) \cdot (x_0 - x_2)}}_{\equiv L_{2,0}(x)} + f(x_1) \cdot \underbrace{\frac{(x - x_0) \cdot (x - x_2)}{(x_1 - x_0) \cdot (x_1 - x_2)}}_{\equiv L_{2,1}(x)} +$$

$x_0 \qquad x_1 \qquad x_2$

$$+ f(x_2) \cdot \underbrace{\frac{(x - x_0) \cdot (x - x_1)}{(x_2 - x_0) \cdot (x_2 - x_1)}}_{\equiv L_{2,2}(x)}$$

The Lagrange polynomial fits a curve to multiple points and therefore improves on the strictly local nature of the Taylor polynomial which is concentrated at a single point.

Linear Interpolation shown in the top figure fits a straight line to two points $(x_0, f(x_0))$ & $(x_1, f(x_1))$. This is easily accomplished by substituting these two points into the equation for a straight line, *viz.*, $y = m\,x + d$, and solving the resulting two equations in two unknowns for the *slope* m and *y-intercept* d. An alternate method is to consider the interpolating polynomial to be the following linear combination of the two 1st degree Lagrange polynomials a(x) and b(x): $P_2(x) = f(x_0).a(x) + f(x_1).b(x)$. We can easily guess the structure of these Lagrange polynomials a(x) and b(x) because $P_2(x_0)$ must yield the nodal value $f(x_0)$, which requires $a(x_0) =1$, $b(x_0) = 0$; similarly, $P_2(x_1) = f(x_1)$ which requires $a(x_1)= 0$, $b(x_1) = 1$. The unique way to satisfy these conditions is to choose

$$a(x) = (x-x_1) / (x_0-x_1) \qquad ; \qquad b(x) = (x-x_0) / (x_1-x_0).$$

Note that this result follows directly by using the four conditions imposed above to solve four equations to find unique slopes and intercepts the linear polynomials a(x) and b(x); however, the intuitive method used above easily generalizes to any number of nodal points without any work!

Quadratic Interpolation shown in the bottom figure fits a quadratic to three points $(x_0, f(x_0))$, $(x_1, f(x_1))$ & $(x_2, f(x_2))$. We write down the linear combination of three 2nd degree polynomials a(x), b(x), and c(x): $P_2(x) = f(x_0)*a(x) + f(x_1)*b(x) + f(x_2) *c(x)$. The requirement now is that a(x) vanishes at both x_1 and x_2 and is equal to unity at x_0 and we immediately write

$$a(x)= [(x-x_1)\ (x-x_2)] / [(x_0-x_1)\ (x_0-x_2)];$$

similar results for b(x) and c(x) follow the same pattern.

3.4.1 Lagrange Interpolating Polynomials - General

Lagrange Interpolating Polynomials - General

- n^{th} degree Expansion

$$P_n^L(x) \equiv \sum_{\alpha=0}^{n} L_{n,\alpha}(x) \cdot f(x_\alpha)$$

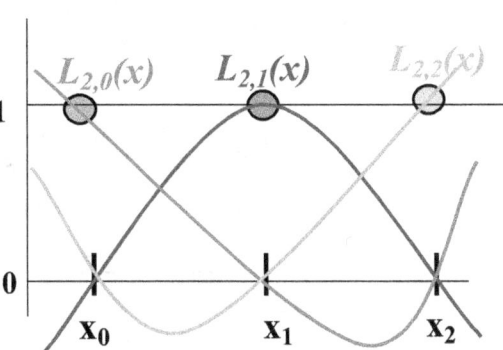

- Lagrange "Basis" Poly

$$L_{n,\alpha}(x) \equiv \frac{(x-x_0)}{(x_\alpha-x_0)} \cdot \frac{(x-x_1)}{(x_\alpha-x_1)} \cdots \overset{\overset{\alpha}{\downarrow}}{\frac{1}{1}} \cdots \frac{(x-x_n)}{(x_\alpha-x_n)}$$

- Note: Different Polys for diff degree n: $L_{n,\alpha}(x)$

The two sets of Lagrange basis polynomials associated with P_1 {a(x), b(x)} and P_2 {a(x), b(x), c(x)} given on the previous slide have nothing in common since the first set is linear and the second set is quadratic. The P_3 and the general P_n shown on this slide are cubic and n^{th} degree polynomials respectively and there is no commonality among these different basis sets except the algorithm by which we generate them.

Thus, given n+1 nodes we generate n+1 basis Lagrange polynomials each of degree n. These polynomials are designated by a pair of indices $L_{n,k}(x)$, where n is the degree of each polynomial and k=0,1,2,...,n indexes the nodal value $f(x_k)$ that multiplies it, $e.g.$, $f(x_k)*L_{n,k}(x)$, in forming the complete Lagrange polynomial.

The general expression for Lagrange basis polynomial $L_{n,k}(x)$ is a product of all $(x-x_j)$ factors j=0,1,..n except that a "1" is put in place of the j=k factor $(x-x_k)$ as illustrated in the slide. This assures us that each $L_{n,k}(x)$ will be zero when evaluated at any node other than $x=x_k$; to make $L_{n,k}(x)=1$ at $x=x_k$, we divide by the product of all (x_k-x_j) factors j=0,1,..n, except the j=k factor (x_k-x_k) is replaced by "1" .

The plot shows the unique quadratic basis polynomials for the 3 points {x_0, x_1, x_2} designated as $L_{2,0}(x)$, $L_{2,1}(x)$, $L_{2,2}(x)$ respectively. The basis polynomial $L_{2,0}(x)$ has the quadratic shape determined by the requirements that it is "1" at x_0 and "0" at both x_1 and x_2. Similar statements apply to the other two basis polynomials $L_{2,1}(x)$, $L_{2,2}(x)$; all three are defined without reference to the nodal values. The nodal values {$f(x_0)$, $f(x_1)$, $f(x_2)$} are just numerical multipliers used to form the full Lagrange polynomial for a specific problem which then goes through {$f(x_0)$, $f(x_1)$, $f(x_2)$} rather than {1,1,1}.

3.4.2 Lagrange Poly: Effects of Adding Points

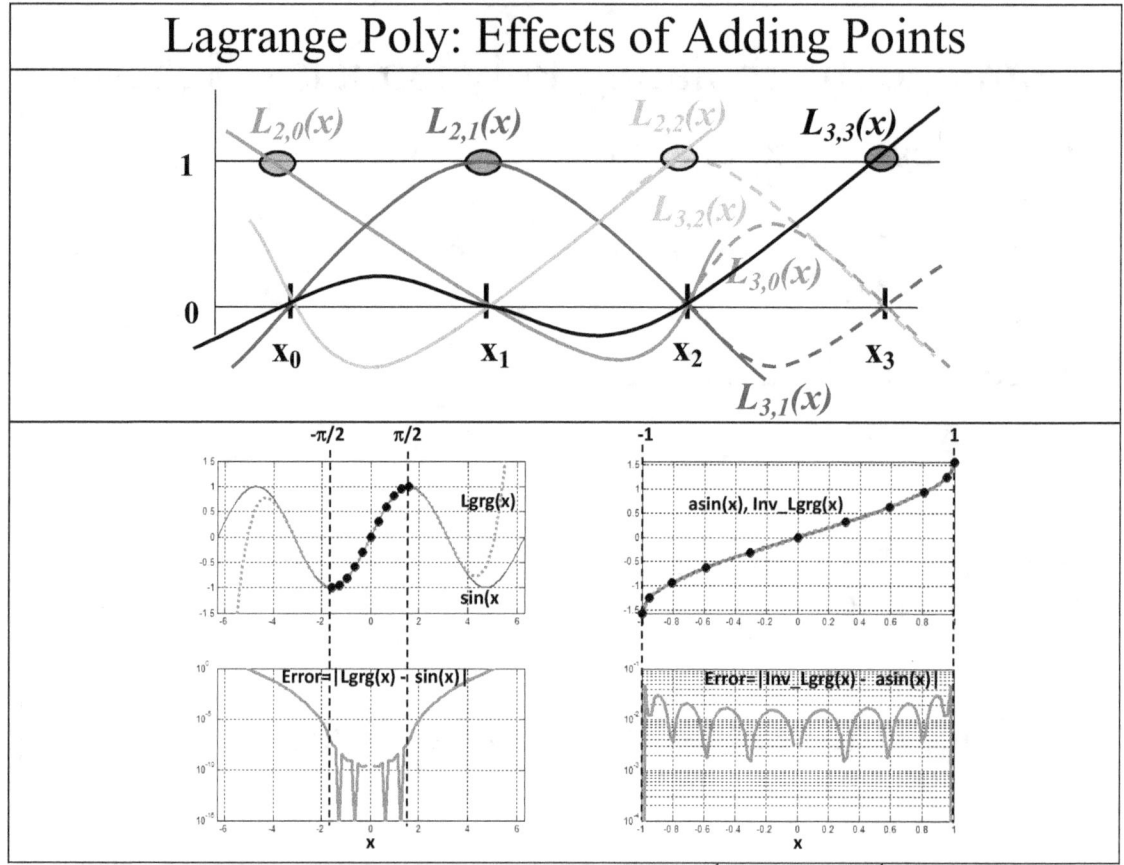

Lagrange Poly: Effects of Adding Points

Upper Panel of this slide illustrates the fact that even though 2nd degree and 3rd degree Lagrange basis polynomials have some nodal points in common, *e.g.*, $f(x_0) = L_{2,0}(x=x_0) = L_{3,0}(x=x_0)$, $f(x_1) = L_{2,1}(x=x_1) = L_{3,1}(x=x_1)$, and $f(x_2) = L_{2,2}(x=x_2) = L_{3,2}(x=x_2)$, they are completely different polynomials. This should be obvious, because they are of different degree and moreover, because their respective basis sets contain 3 and 4 distinct polynomials. The figure further emphasizes the increased "oscillation" as the degree increases from 2 to 3. For example, the blue curve showing $L_{2,1}(x)$ continues down and never re-crosses the x-axis, whereas $L_{3,1}(x)$ is "forced " to turn up again so as to go through "0" at the new node x_3 as shown by the dashed extension of the blue curve. Similar changes occur for $L_{3,0}(x)$ (red) and $L_{3,2}(x)$ (green) which now need to turn in order to go through "0" at the new node x_3. Moreover, a new basis polynomial $L_{3,3}(x)$ (black) is required to go through "1" at the 4th node x_3 and through "0" at the other three nodes.

Lower Panel shows that the oscillation problem not withstanding, additional points can often give a very accurate representation of a smooth function over a large interval; this should be compared to the results of Slide#3-10 where the accuracy of a Taylor polynomial is limited to a small region about the evaluation point x_0. The plots on the left compare $\sin(x)$ (solid, blue) and the Lagrange polynomial Lgrg(x) (dashed red curve) over the interval $[-2\pi, 2\pi]$ and below is the absolute difference on a logarithmic scale. The Lagrange polynomial uses the 11 sample points of the sine function in $[-\pi/2, \pi/2]$ (black dots within dashed vertical lines). The error is well below 10^{-5} in $[-\pi/2, \pi/2]$ and remains around $\sim 10^{-2}$ in $[-4, 4]$. The dips in the error curve correspond to the nodal points (black dots). Similar remarks hold for the inverse Lagrangian and asin(x), but the errors are much larger ranging 3×10^{-3} to 5×10^{-2} near the ends. This is somewhat intuitive since the inverse is constrained to be within the invertible interval $[-1,1]$ and hence there is no place else to dump the errors; in the direct case, periodicity places ("aliases") some error outside the sample interval $[-\pi/2, \pi/2]$ (MatLab Script on Slides #5-7,5-8,5-9).

3.4.3 Example: Linear Combination of Basis Polynomials

Example:Lin. Comb. of Basis Polynomials

Linear (2 pts)

$$P_1^L(x) = 6 \cdot \frac{x-1}{0-1} + 1 \cdot \frac{x-0}{1-0}$$

$$= 6 \cdot \underbrace{(1-x)}_{L_{1,0}(x)} + 1 \cdot \underbrace{x}_{L_{1,1}(x)}$$

$$= 6 - 5x$$

Nodal Table	
x	**f(x)**
0	6
1	1
2	3

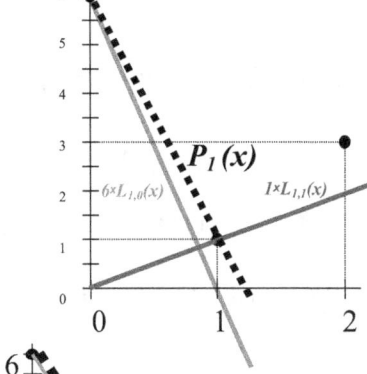

Quadratic(3 pts)

$$P_2(x) = 6 \cdot \frac{(x-1)\cdot(x-2)}{(0-1)\cdot(0-2)} + 1 \cdot \frac{(x-0)\cdot(x-2)}{(1-0)\cdot(1-2)} + 3 \cdot \frac{(x-0)\cdot(x-1)}{(2-0)\cdot(2-1)}$$

$$= 6 \cdot \underbrace{\frac{1}{2}(x^2 - 3x + 2)}_{L_{2,0}(x)} + 1 \cdot \underbrace{(2x - x^2)}_{L_{2,1}(x)} + 3 \cdot \underbrace{\frac{1}{2}(x^2 - x)}_{L_{2,2}(x)}$$

$$= \frac{7x^2 - 17x + 12}{2} = \frac{1}{2} \cdot [x \cdot (7 \cdot x + 17) + 12]$$

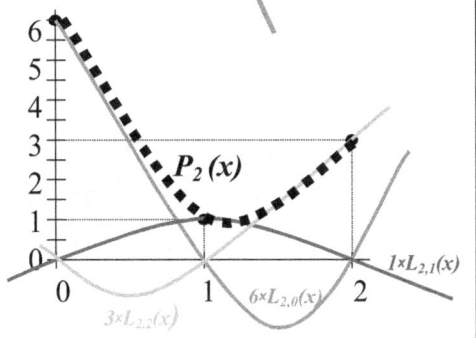

Here are two explicit examples using 2 and 3 points from the same nodal table.

Linear: Using the first two nodal points we get one possible linear Lagrange expansion consisting of the two straight lines shown in the top figure. Each straight line is multiplied by its nodal value 6* $L_{1,0}(x)$ and 1 * $L_{1,1}(x)$ and then summed to yield the dashed interpolating straight line. We could of course use the last two points and get a different Lagrange expansion.

Quadratic: Using the all three nodal points, we get the quadratic Lagrange expansion consisting of the three quadratic curves shown in the bottom figure. Each quadratic curve is multiplied by its nodal value, *viz.*, 6* $L_{2,0}(x)$, 1 * $L_{2,1}(x)$, 3 * $L_{2,2}(x)$, and then summed to yield the dashed interpolating quadratic curve.

The Lagrange polynomial is written down in several forms. The factored form given in the 1st equality is easily written down directly from the table and is also easily checked to make certain that each Lagrange basis polynomial reduces to "1" at the appropriate node $x = x_k$ and to "0" at the other nodes. The second equality shows the explicit basis polynomials and their multipliers, and moreover, makes clear the interpretation of the Lagrange polynomial expansion as a simple "scaling" of each basis function $L_{n,k}(x)$, by its nodal value $f(x_k)$ before taking the sum. Further algebraic simplification yields the quadratic form in the third equality, and finally nesting that expression yields the most efficient numerical form with 3 multiplications and 2 adds (quadratic case). However, for such a low degree polynomial, it is OK to just evaluate the first form which is less prone to human errors.

3.4.4 Creating Table With Specified Error

Creating Table With Specified Error

- Lagrange format - works with numbers directly
- Look-up table for function 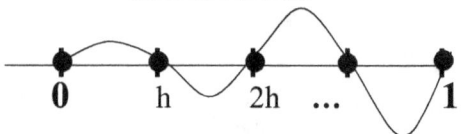 $f(x) = e^x \ ; \ x \ \varepsilon \ [0,1]$
- Find required stepsize $h = \dfrac{1-0}{N}$
  - Linear Interpolation
  - Error $\varepsilon < 10^{-d}$

Error zero at nodes

$$0 \qquad h \qquad 2h \quad \ldots \qquad 1$$

- Results $N = \sqrt{\dfrac{e \cdot 10^d}{8}}$

  or $\quad h = \dfrac{1}{N}$

Error (ε)	Stepsize (h)	Entries (N)
10^{-6}	.0017 ➔ .001	583 ➔ 1000
10^{-3}	.054 ➔ .05	19 ➔ 20
10^{-2}	.17 ➔ .1	6 ➔ 10

In some computational situations it may be convenient to use a simple look up table instead of calculating the function directly for each new input value. In order for an interpolation table to be a useful substitute for the actual function, it must guarantee that the interpolations between the tabulated values (nodes) achieve the desired accuracy be it 10^{-2} or 10^{-3} or in general $\varepsilon_{Tol.} = 10^{-d}$. It must also contain a reasonable number of points since we must always trade off computation against storage and input/output costs.

This error tolerance is achieved across the entire table by judiciously choosing an appropriate step size and number of steps N. Since the Lagrange interpolating polynomial is constructed so as to yield the exact table values at the nodes, we only need worry about exceeding our error tolerance criterion between the nodes as shown in the illustration.

The result (see next Slide# 3-16) for linear interpolation of an exponential function exp(x) over the range [0, 1] is given by a formula that expresses the number of table entries N needed to stay within the specified tolerance $\varepsilon_{Tol.} = 10^{-d}$. The short table evaluates N exactly and extends them (➔) to convenient values for a few error tolerances $\varepsilon_{Tol.} = 10^{-d}$; for 10^{-2} maximum error, we need N=10 table entries, while for 10^{-3} maximum error we need 20, and for 10^{-6} maximum error, we need a whopping 1000 table entries. Clearly there must be a trade-off between the storage requirements of a large table versus the cost of computing the function directly.

Clearly, for 10^{-6} error tolerance, a look up table containing 1000 entries may not be a good choice relative to direct computation; it may be that we should look at higher degree interpolation. On the next slide we derive this result using the explicit error formula for linear Lagrange interpolation.

3.4.5 Lagrange Interpolation Table - Details

Lagrange Interpolation Table - Details

- Table error for Lagrange Interpolation degree n=1
- Consider nodes x_0 & $x_1 = x_0 + h$ and let $y = x - x_0$, so that $y \, \varepsilon \, [0,h]$

$$\varepsilon(x,\xi) = \left| R_1^L(x,\xi) \right| = \max_{x,\xi} \left| \frac{(x-x_0) \cdot (x-x_1)}{2} \cdot f^{(2)}(\xi) \right|$$

Error zero at nodes

$$= \max_{intvls} \frac{y(y-h)}{2} \cdot \max_{\xi \in [0,1]} e^{\xi} < 10^{-6}$$

$$\underbrace{\phantom{\frac{y(y-h)}{2}}}_{\equiv g(y)}$$

- Same g(y) for all intervals

$$g'(y) = \frac{2y-h}{2} = 0 \;\Rightarrow\; y_{max} = h/2 \qquad \varepsilon = \left| \frac{h/2(h/2 - h)}{2} \right| \cdot e^1 < 10^{-d}$$

- Required stepsize h (number of entries $N = 1/h$) is

$$\frac{eh^2}{8} = \frac{e}{8N^2} < 10^{-d} \;\Rightarrow\; N = \sqrt{\frac{e \cdot 10^d}{8}}$$

For linear interpolation, the truncation error is quadratic and the Lagrange error for any x between a pair of nodes, *e.g.*, x_0, x_1 is given by the maximum of [1/2 $(x-x_0)$ $(x-x_1)$ $f^{(2)}(\xi)$] for all x and ξ in the interval $[x_0, x_1]$.

Since we want this error to be less than 10^{-d} everywhere in [0, 1], we should really just look at the last nodal interval $[x_{n-1}, x_n]$, which has $x_n = 1$ because that is where the 2nd derivative e^{ξ} is largest ($\xi = 1$) *i.e.*, $\max[f^{(2)}(\xi)] = e^1$.

The product term can be transformed into a form g(y)=y(y-h)/2, which is the same for any nodal interval and leads to a maximum y_{max}=h/2. Substitution back into the error expression yields the upper bound on the error as a function of the step size h; finally recognizing that the number of steps N=(1-0)/h we can solve for the N required to cover the whole interval [0,1] with the desired accuracy.

3.4.6 Choosing Degree of Lagrange Polynomial

Choosing Degree of Lagrange Polynomial

- How do we choose optimal degree of polynomial?
- Higher degree not always better!
- Add more points (higher degree)
- Iteration scheme on polynomial *degree n*

  Yields new polynomial of *degree (n+1) - efficient*
- Use a stopping criteria ("cf. diagonal terms")
- Tableau row-by-row addition of points.

If we have a nodal table with say 8 points $\{x_0, x_1,..., x_7\}$, then by taking various combinations of points we could write down many different Lagrange interpolation polynomials with degrees ranging from 1 to 7. They all need to be calculated individually since even if they are the same degree, say 3, the choice of different sets of 4 points yields completely different polynomials.

If we want to find an estimate of the underlying function $f(x)$ at a fixed value, say, $x = 1.5$, then the question is which nodes and what degree Lagrange polynomial yields the best estimate of $f(1.5)$. The Lagrange polynomial based on the nodes closest to $x=1.5$ clearly are the best candidates since they will "capture" the functional variation of the table near $x=1.5$. However, although increasing the degree of the polynomial to 4, 5, 6, or 7 will account for more of the latent global structure of the function, it is not clear how to proceed or when to stop. Higher degree may or may not yield better results since higher degree introduces more oscillations by virtue of the requirement that each basis function be zero at all the nodes except for the one to which it corresponds. This whole process is computationally expensive.

Iterated interpolation is an efficient method for generating higher degree Lagrange polynomials from lower degree ones by simply adding rows corresponding to new nodes. It uses a very convenient tableau similar to Horner's method and generates all possible Lagrange interpolation polynomials for pairs, triples, and so on. Moreover, comparing the magnitudes of diagonal terms as we proceed row-by-row allows for a logical stopping criterion as their difference becomes small.

3.5 *Iterated Interpolation Tableau*

Iterated Interpolation Tableau

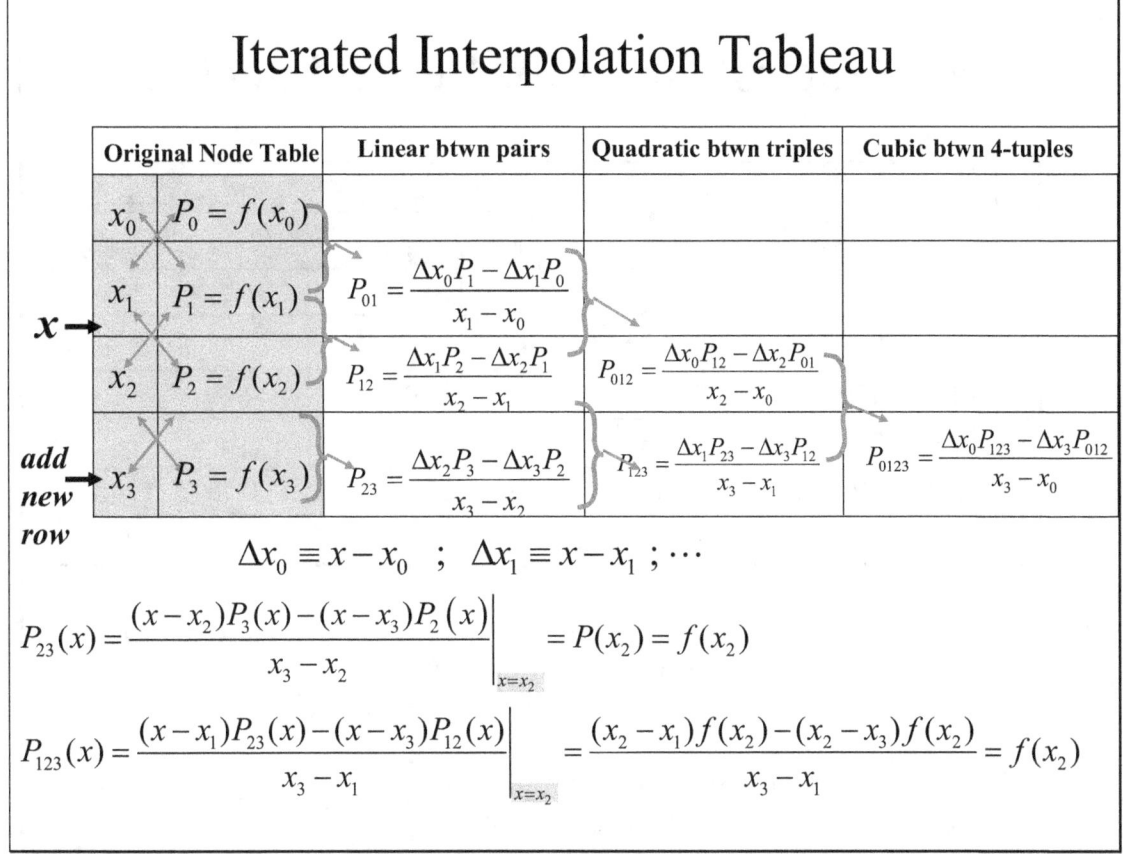

Original Node Table		Linear btwn pairs	Quadratic btwn triples	Cubic btwn 4-tuples
x_0	$P_0 = f(x_0)$			
x_1	$P_1 = f(x_1)$	$P_{01} = \dfrac{\Delta x_0 P_1 - \Delta x_1 P_0}{x_1 - x_0}$		
x_2	$P_2 = f(x_2)$	$P_{12} = \dfrac{\Delta x_1 P_2 - \Delta x_2 P_1}{x_2 - x_1}$	$P_{012} = \dfrac{\Delta x_0 P_{12} - \Delta x_2 P_{01}}{x_2 - x_0}$	
add x_3 *new* *row*	$P_3 = f(x_3)$	$P_{23} = \dfrac{\Delta x_2 P_3 - \Delta x_3 P_2}{x_3 - x_2}$	$P_{123} = \dfrac{\Delta x_1 P_{23} - \Delta x_3 P_{12}}{x_3 - x_1}$	$P_{0123} = \dfrac{\Delta x_0 P_{123} - \Delta x_3 P_{012}}{x_3 - x_0}$

$$\Delta x_0 \equiv x - x_0 \quad ; \quad \Delta x_1 \equiv x - x_1 \; ; \; \cdots$$

$$P_{23}(x) = \frac{(x - x_2)P_3(x) - (x - x_3)P_2(x)}{x_3 - x_2}\bigg|_{x = x_2} = P(x_2) = f(x_2)$$

$$P_{123}(x) = \frac{(x - x_1)P_{23}(x) - (x - x_3)P_{12}(x)}{x_3 - x_1}\bigg|_{x = x_2} = \frac{(x_2 - x_1)f(x_2) - (x_2 - x_3)f(x_2)}{x_3 - x_1} = f(x_2)$$

We shall first set up the iterated interpolation tableau and then verify why and how it works. The tableau does not compute the Lagrange basis functions $L_{n,k}(x)$ corresponding to each degree n and labeled nodes k =0,1,2,....,n. Instead, the tableau directly computes the full Lagrange polynomial P_1, P_2, P_3 *etc.*. Because we need to distinguish between various choices of 2, of 3, of 4 nodal points we introduce a new notation for which the subscripts label the specific nodes used to construct the Lagrange polynomial; the degree is implicit since it is just the number of labeled nodes minus 1.

Thus $P_{01}(x)$ has two nodes and it is a (2-1) 1st degree Lagrange polynomial which passes through the nodes x_0 and x_1; more specifically, this means that $P_{01}(x=x_0)=f(x_0)$ and $P_{01}(x=x_1)=f(x_1)$. Similarly, $P_{123}(x)$ has three nodes and it is (3-1) 2nd degree and takes on the nodal values at x_0, x_1, and x_2, *etc.*. For notational conformity we replace the nodal value $f(x_0)$ with the "number" P_0; note that this is consistent since the single index implies a polynomial of degree 1-1=0 which is just a constant or single number.

The nodal table is "entered" with a value x(=1.5) and the term $P_{01}(x=1.5)$ is obtained by cross multiplying $[(x-x_0)P_1 - (x-x_1)P_0]$ and dividing by the node difference (x_1-x_0). Similarly, for the term $P_{12}(x=1.5)$ and the term $P_{23}(x=1.5)$ to complete the Linear column between pairs of points. Next the term $P_{012}(x=1.5)$ is obtained by cross multiplying $[(x-x_0)P_{12} - (x-x_2)P_{01}]$ and dividing by the node difference (x_2-x_0). Similarly, for the term $P_{123}(x=1.5)$ to complete the Quadratic column between triples. Finally, the term $P_{0123}(x=1.5)$ is obtained by cross multiplying $[(x-x_0)P_{123}-(x-x_3)P_{012}]$ and dividing by the node difference (x_3-x_0). In this way all linear and all quadratic and the one cubic Lagrange polynomial evaluations at the point x=1.5 have been produced in a single table. Additional points (rows) may be added and processed in a similar manner.

3.5.1 Iterated Interpolation for Bessel Function

Iterated Interpolation for Bessel Function

				Bessel Fcn 1st kind			
k	x_k	$f(x_k)=P_k$	P_{ab}	P_{abc}	P_{abcd}	P_{abcde}	P_{abcdef}
0	1.0	.7651977=P_0	P_{01}				
1	1.3	.6200860=P_1	.5233449	P_{012}			
2	1.6	.4554022=P_2	.5102968	.5124715	P_{0123}		
3	1.9	.2818186	.5132634	.5112857	.5118127	P_{01234}	
4	2.2	.1103623	.5104270	.5137361	.5118302	.5118200	P_{012345}
5	2.5	-.0483838	.4807699	...	...	...	.5118277

$x=1.5$

Sample Calc.

$$P_{01} = \frac{\Delta x_0 P_1 - \Delta x_1 P_0}{x_1 - x_0} = \frac{\overbrace{(1.5-1.0)}^{\Delta x_0} \cdot \overbrace{(.6200860)}^{P_1} - \overbrace{(1.5-1.3)}^{\Delta x_1} \cdot \overbrace{(.7651977)}^{P_0}}{1.3-1.0} = .5233449$$

$$P_{012} = \frac{\Delta x_0 P_{12} - \Delta x_2 P_{01}}{x_2 - x_0} = \frac{\overbrace{(1.5-1.0)}^{\Delta x_0} \cdot \overbrace{(.5102968)}^{P_{12}} - \overbrace{(1.5-1.6)}^{\Delta x_2} \cdot \overbrace{(.5233449)}^{P_{01}}}{1.6-1.0} = .5124715$$

Exact:

$b_1(1.5) = .5118277$

The iterated interpolation process is applied to the 6 rows of the tabulated Bessel Function. The numbers in this table correspond to the entry value x=1.5 and result from the sequential cross multiplication of "Δx" s and "P" s and division as per the template on the last slide. The interpolated values are tabulated for all linear, all quadratic, all cubic, all quartic, all quintic, and the one polynomial using all 6 points. Two sample calculations are given below the table.

After completing each row we can compare the magnitude of the successive diagonal terms to see how the sequence is converging. If we only need 3 sd, we would stop at row 3 or 4 since the difference between diagonal terms is of order 10^{-4}, or we can continue to the end of the table and obtain 7 significant digits. In this way the iterated interpolation tableau method answers our original question (Slide#3-17) on how to choose the degree of the Lagrange Polynomial *via* a numerical stopping criterion, *i.e.*, compare the diagonal terms.

3.6 *Newton Divided Difference Polynomials*

Newton Divided Difference Polynomials

- ## *Lagrange Interpolation Tables*
  - Displays interpolated function value
  - Evaluate *f(x)* at a specific point *x (=1.5)*
  - Can iterate to refine the evaluation
  - *New table* needed for a *different point*
- ## *Divided Difference Tables*
  - Displays polynomial coefficients
  - Single Interpolating polynomial
  - Subsequently evaluated for many "nearby" points

$$P_n^F(x) = \sum_{k=0}^{n} f[x_0 x_1 \cdots x_k] \cdot \underbrace{(x-x_0) \cdot (x-x_1) \cdots (x-x_{k-1})}_{k-factors}$$

The iterated interpolation table generates an interpolated value for a single point (x=1.5 in the example). If we choose a new point x=1.7, then we need to compute a new table to find the interpolated value. We could of course generate the Lagrange polynomials using this iterative technique by simply leaving the value of x unspecified, but then the table would involve algebraic manipulations and lose its "numerical simplicity." Moreover, there would be no numerical stopping criterion since everything is done algebraically.

Thus, iterated interpolation is an evaluation of the successive Lagrange polynomials at a fixed x-value. It would be nice to have a numerical method for generating the **coefficients** rather than the **interpolated values** of successive polynomials.

The Newtonian Divided Difference table is such a procedure; it uses the same nodal table and generates the coefficients of the interpolating polynomial; new polynomials are obtained by adding new nodes (rows) to the table.

The equation shows the basic equation defining the Newtonian polynomial as a sum of terms containing k^{th} degree polynomials $(x-x_0)(x-x_1) \ldots (x-x_k)$ multiplied by the divided difference coefficients $f[x_0 x_1 \ldots x_k]$ which are generated in the DD table shown on the next page. For example the n=2 Newtonian polynomial is written as
$P_2^F(x) = f[x_0] + f[x_0 x_1](x-x_0) + f[x_0 x_1 x_2](x-x_0)(x-x_1)$
The coefficients are obtained by requiring that the polynomial reduce to the nodal values at the three nodes $\{x_0, x_1, x_2\}$; thus at $x=x_0$, we have $f(x_0) = P_2(x=x_0) = f[x_0] + 0 + 0$; at $x=x_1$, we have $f(x_1) = P_2(x=x_1) = f[x_0] + f[x_0 x_1](x_1-x_0) + 0$ which yields the 1st DD coefficient $f[x_0 x_1] = [f(x_1)-f(x_0)] / (x_1-x_0)$; continuing in this fashion yields the 2nd DD coefficient $f[x_0 x_1 x_2] = \{f[x_1 x_2] - f[x_0 x_1]\} / (x_2-x_0)$. The table on the next slide generalizes these DD coefficients in a natural way.

3.6.1 Divided Difference Table

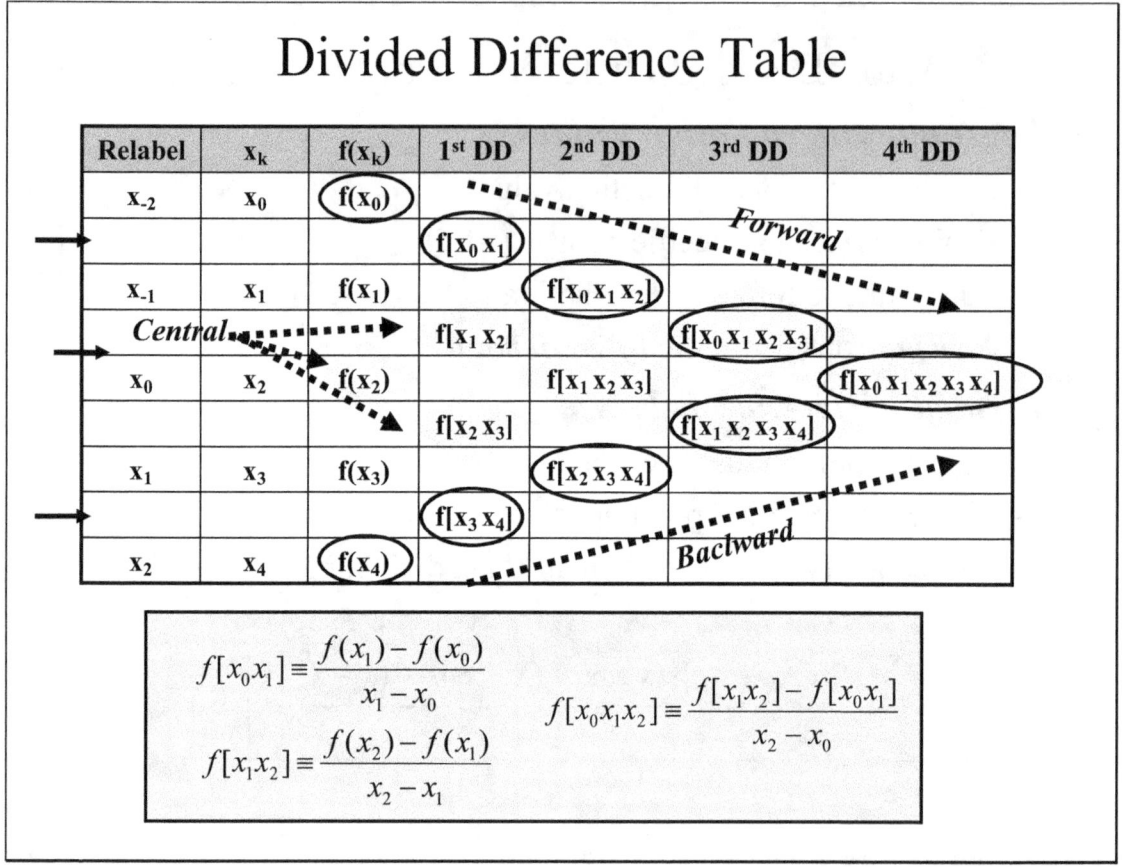

Divided Difference Table

Relabel	x_k	$f(x_k)$	1ˢᵗ DD	2ⁿᵈ DD	3ʳᵈ DD	4ᵗʰ DD
x_{-2}	x_0	$f(x_0)$				
			$f[x_0 x_1]$			
x_{-1}	x_1	$f(x_1)$		$f[x_0 x_1 x_2]$		
Central			$f[x_1 x_2]$		$f[x_0 x_1 x_2 x_3]$	
x_0	x_2	$f(x_2)$		$f[x_1 x_2 x_3]$		$f[x_0 x_1 x_2 x_3 x_4]$
			$f[x_2 x_3]$		$f[x_1 x_2 x_3 x_4]$	
x_1	x_3	$f(x_3)$		$f[x_2 x_3 x_4]$		
			$f[x_3 x_4]$			
x_2	x_4	$f(x_4)$				

$$f[x_0 x_1] \equiv \frac{f(x_1) - f(x_0)}{x_1 - x_0}$$

$$f[x_1 x_2] \equiv \frac{f(x_2) - f(x_1)}{x_2 - x_1}$$

$$f[x_0 x_1 x_2] \equiv \frac{f[x_1 x_2] - f[x_0 x_1]}{x_2 - x_0}$$

This single table generates the DD coefficients for three different Newtonian polynomials, namely **forward** (F), **backward** (B) and **central** (C). Because a polynomial is uniquely determined by its nodal values, it might be thought that there can be no difference between these three polynomials. This would be true if we used "exact" arithmetic, but in fact we use finite digit arithmetic. So what is the difference? The numerical inaccuracies of finite digit arithmetic will make the choice of polynomial hinge upon the location of the x "entry value"; thus we choose forward coefficients for x near the top, backward for x near the bottom, and central for x-value in the center.

3.6.2 Forward, Backward, and Central Polynomials

Forward, Backward, and Central Polynomials

x_k	$f(x_k)$	1st DD	2nd DD	3rd DD	4th DD
1.0	.7651977				
		-.4837057			
1.3	.6200860		-.1087339		
		-.5489460		.0658784	
1.6	.4554022		-.0494433		.0018251
		-.5786120		.0680685	
1.9	.2818186		+.0118183		
		-.5715210			
2.2	.1103623				

$P_4^{Fwd}(x) = .7651977 - .4837057 \cdot (x-1.0) - .1087339 \cdot (x-1.0) \cdot (x-1.3)$
$+ .0658784 \cdot (x-1.0) \cdot (x-1.3)(x-1.6) + .0018251 \cdot (x-1.0) \cdot (x-1.3)(x-1.6)(x-1.9)$

$P_4^{Bkwd}(x) = .1103623 - .5715210 \cdot (x-2.2) + .0118183 \cdot (x-2.2) \cdot (x-1.9)$
$+ .0680685 \cdot (x-2.2) \cdot (x-1.9)(x-1.6) + .0018251 \cdot (x-2.2) \cdot (x-1.9)(x-1.6)(x-1.3)$

$P_4^{Cntrl}(x) = .4554022 + \dfrac{-.5489460 - .5786120}{2} \cdot (x-1.6) - .0494433 \cdot \dfrac{(x-1.6)(x-1.9) + (x-1.3)(x-1.6)}{2}$
$+ \dfrac{.0658784 + .0680685}{2} \cdot (x-1.3) \cdot (x-1.6) \cdot (x-1.9) + .0018251 \cdot \dfrac{(x-1.3)(x-1.6)(x-1.9)(x-2.2) + (x-1.0)(x-1.3)(x-1.6)(x-1.9)}{2}$

This example shows the complete DD table generated from the initial data table (5-nodes). The resulting forward, central, and backward polynomials suitable for different x entry values (indicated by the black arrows at the left edge of the table) are written down by "taking" the required coefficients off the table and placing them in the appropriate form for each as follows:

Forward: $P_4^F(x) = f[x_0] + f[x_0x_1] (x-x_0) + f[x_0x_1x_2] (x-x_0)(x-x_1)$
$+ f[x_0x_1x_2x_3] (x-x_0)(x-x_1)(x-x_2) + f[x_0x_1x_2x_3x_4] (x-x_0)(x-x_1)(x-x_2) (x-x_3)$

Backward: $P_4^B(x) = f[x_4] + f[x_3x_4] (x-x_4) + f[x_2x_3x_4] (x-x_4)(x-x_3)$
$+ f[x_1x_2x_3x_4] (x-x_4)(x-x_3)(x-x_2) + f[x_0x_1x_2x_3x_4] (x-x_4)(x-x_3)(x-x_2) (x-x_1)$

Re-label points: $\{x_0x_1x_2x_3x_4\} \rightarrow \{x_{-2}x_{-1}x_0x_1x_2\}$
Central: $P_4^C(x) = f[x_0] + \frac{1}{2} \{f[x_{-1}x_0] + f[x_0x_1]\}(x-x_0)$
$+ f[x_{-1}x_0x_1] \{(x-x_{-1})(x-x_0) + (x-x_0)(x-x_1)\}/2 +$
$+ \frac{1}{2} \{f[x_{-2}x_{-1}x_0x_1] + f[x_{-1}x_0x_1x_2]\}(x-x_{-1})(x-x_0)(x-x_1)$
$+ f[x_{-2}x_{-1}x_0x_1x_2] \{(x-x_{-2}) (x-x_{-1})(x-x_0) (x-x_1) + (x-x_{-1})(x-x_0)(x-x_1) (x-x_2) \}/2$

"hopscotch" pattern: after first term alternately averaging surrounding polynomials, then averaging nodal values about new x_0"

3.6.3 Inverse Newton Interpolation Polynomials

Inverse Newton Interpolation Polynomials

$y_k=f(x_k)$	$x_k=g(y_k)$	1st DD	2nd DD	3rd DD	4th DD
.7651977	1.0				
		-2.067373			
.6200860	1.3		.79310384		
		-1.821673		-1.069540	
.4554022	1.6		-.2761100		-0.61990
		-1.728274		-0.663610	
.2818186	1.9		+.0621500		
		-1.749717			
.1103623	2.2				

$$P_2^{Fwd}(y) = 1.0 - 2.607373 \cdot (y-.7651977) - .79310384 \cdot (y-.7651977) \cdot (y-.6200860)$$

$$P_2^{Bkwd}(y) = 2.2 - 1.749717 \cdot (y-.1103623) + .06215(y-.1103623) \cdot (y-.2818186)$$

$$P_2^{Cntrl}(y) = 1.6 + \frac{-1.821673 - 1.728274}{2} \cdot (y-.4554022)$$
$$-.2761100 \cdot \frac{(y-.4554022)(y-.620086) + (y-.4554022)(y-.2818186)}{2}$$

By swapping the columns in the original table and re-labeling them y_k and $g(y_k)$ respectively, the three Newtonian polynomials generated by this table represent inverse interpolation of the original table. That is, we enter this new table with a f(x)-value "y" and find the interpolated x-value "g(y)". The procedure is identical to the previous slide and generates polynomials in the variable y as shown.

Note that the original table had *equal step sizes* $\Delta x = h = 0.3$ while the new table has *unequal step sizes* of (.6200860-.7651977) , (.4554022-.6200860), (.2818186- .4554022) and (.1103623-.4554022), but this does not affect the algorithm in any way.

3.7 *Uniformly Spaced Difference Tables*

Uniformly Spaced Difference Tables

- ## Uniformly spaced data

$$x = x_0 + s\,h$$

$$\begin{array}{cccc} x_0 & x_1 & x_2 \ldots & x_k \end{array}$$

  – Forward

$$P_n^F(x) = \sum_{k=0}^{n} \underbrace{f[x_0 x_1 \cdots x_k]}_{=\Delta^k f(x_0)\big/ k! h^k} \cdot \underbrace{(x-x_0)}_{sh} \cdot \underbrace{(x-x_1)}_{(s-1)h} \cdots \underbrace{(x-x_{k-1})}_{(s-k+1)h}$$

$$\underbrace{\phantom{(x-x_0)(x-x_1)\cdots(x-x_{k-1})}}_{k-factors}$$

$$\xrightarrow{\;x=x_0+s\cdot h\;} \boxed{\sum_{k=0}^{n} \binom{s}{k} \Delta^k f(x_0)}$$

$$x = x_n + s\,h$$

$$\begin{array}{cccc} x_0 & x_1 & \ldots\; x_{n-2}\; x_{n-1} & \boxed{x_n} \end{array}$$

  – Backward

$$P_n^B(x) = \sum_{k=0}^{n} f[x_n x_{n-1} \cdots x_{n-k}] \cdot \underbrace{(x-x_n)}_{(-s)h} \cdot \underbrace{(x-x_{n-1})}_{((-s)-1)h} \cdots \underbrace{(x-x_{n-k+1})}_{((-s)-k+1)h}$$

$$\underbrace{\phantom{(x-x_n)(x-x_{n-1})\cdots(x-x_{n-k+1})}}_{k-factors}$$

$$\xrightarrow{\;x=x_n+s\cdot h\;} \boxed{\sum_{k=0}^{n} \binom{-s}{k} (-1)^k \nabla^k f(x_n)}$$

If the step size has a fixed value h, then we can incorporate the division part of the DD method by reformulating the problem in a manner that explicitly takes into account the fixed step size h.

Making a coordinate transformation to a new variable "s" which measures distances from x_0 *forward* to a point $x=x_0+sh$ (s arbitrary) produces the terms in the forward polynomial sum shown in the 1st figure. The resulting expression in terms of the variable s is $_{given}$ in the 1st boxed equation in terms of tabulated functional differences (not divided differences) and does not explicitly include the step size h. After writing out the s-polynomial $P_n^F(s)$ directly using coefficients from the Delta table, the step size h is re-introduced by setting $s=(x-x_0)/h$.

Alternately, making a coordinate transformation to a new variable "s," which measures distances from x_n *backward* to a point $x=x_n+sh$ (s arbitrary), generates the terms of the backward polynomial sum shown in the 2nd figure. The resulting expression in terms of the variable s is given in the 2nd boxed equation. After writing out the s-polynomial $P_n^B(s)$ directly using coefficients from the Δ-table, the step size h is re-introduced by setting $s=(x-x_n)/h$.

Note: Equating $f[x_0,x_1,\ldots,x_k] = \Delta^k f(x_0)/(k!\ h^k)$ may be proved by verifying it for the cases k=1,2,3 to establish the pattern or more formally by mathematical induction. Thus we verify the first few cases

k=1: $f[x_0,x_1] = \{f[x_1]-f[x_0]\}/(x_1-x_0) = \{f(x_1)-f(x_0)\}/h = \Delta f(x_0)/h$;

k=2: $f[x_0,x_1,x_2] = \{f[x_1,x_2] - f[x_0,x_1]\}/(x_2-x_0) = \{\Delta f(x_1)/h - \Delta f(x_0)/h\ \}/(2h) = \Delta^2 f(x_0)/(2!\ h^2)$

[where above, we have used the fact that $\Delta^2 f(x_0) = \Delta\{\Delta f(x_0)\} = \Delta\{f(x_1)-f(x_0)\} = \{\Delta f(x_1) - \Delta f(x_0)\}$]

k=3: $f[x_0,x_1,x_2,x_3] = \{f[x_1,x_2,x_3] - f[x_0,x_1,x_2]\}/(x_3-x_0) = \{\Delta^2 f(x_1)/2h^2 - \Delta^2 f(x_0)/2h^2\}/\ 3h = \Delta^3 f(x_0)/(3!\ h^3)$

3.7.1 Physics Lab Spark Strip & "g"

Physics Lab Spark Strip & "g"

- Drop metal ball from rest and measure distance btwn spark marks
- Table of 1st and 2nd differences Δx , $\Delta^2 x$

t	x(t)	Δx	Δ²x
0	100		
		-4.9	
1	95.1		-9.8
		-14.7	
2	80.4		-9.8
		-24.5	
3	55.9		-9.8
		-34.3	
4	21.6		

g = -9.8

Note: The DD table for this problem looks almost identical except $\Delta^2 x$ col is divided by $t_2 - t_0 = 2$

$$P_2^F(t) = \sum_{k=0}^{2} \binom{s}{k} \Delta^k x(t_0) \qquad \text{Forward}$$

$$= \binom{s}{0} \Delta^0 x(t_0) + \binom{s}{1} \Delta^1 x(t_0) + \binom{s}{2} \Delta^2 x(t_0)$$

$$= 100 + s(-4.9) + \frac{s(s-1)}{2}(-9.8)$$

$$= 100 - 4.9 s^2 \xrightarrow{s=t} 100 - 4.9 t^2$$

$$P_2^B(t) = \sum_{k=0}^{2} \binom{-s}{k}(-1)^k \nabla^k f(x_n) \qquad \text{Backward}$$

$$= \binom{-s}{0}(-1)^0 \nabla^0 x(t_2) + \binom{-s}{1}(-1)^1 \nabla^1 x(t_2) + \binom{-s}{2}(-1)^2 \nabla^2 x(t_2)$$

$$= 80.4 + (-s)(-1)(-14.7) + \frac{-s(-s-1)}{2}(-1)^2(-9.8)$$

$$= 80.4 - 19.6 s - 4.9 s^2 \xrightarrow{s=t-2} 100 - 4.9 t^2$$

The data table shows the position versus time for an object in free fall. The table data has a constant step size of $\Delta t = 1$ and so we simply form a difference table using the first three points. The coefficients for the forward and backward differences are circled and end in a common one -9.8. The two quadratic polynomials $P^F_2(s)$ and $P^B_2(s)$ each have three terms that are explicitly written out from their defining formulas and it is noted that they lead to different polynomials in the variable s. However, applying the two different coordinate transformations yields the same result in the time variable t.

It should be noted that the binomial coefficient with the $-s$, e.g., $^{-s}C_2$ can be obtained by first computing $^s C_2 = s(s-1)/2$ and then letting $s \to -s$ to yield $(-s)(-s-1)/2$; also take care to note the additional sign changing term $(-1)^k$ in the backward polynomial.

3.8 *Hermite Interpolating Polynomials*

Hermite Interpolating Polynomials

- ## Requires Elaborate Computations

Time	Pos	Vel
x_k	$f(x_k)$	$f'(x_k)$
5.0	2.168	-1.495
5.2	1.797	-1.266
5.4	1.488	-1.070

$$f(x) \approx H_{2n+1}(x) \equiv \sum_{k=0}^{n} \left(\underbrace{F_{n,k}(x) \cdot f(x_k)}_{\text{nodal vals.}} + \underbrace{D_{n,k}(x) \cdot f'(x_k)}_{\text{nodal derivs.}} \right)$$

$$F_{n,k}(x) \equiv \left[1 - 2(x - x_k) L'_{n,k}(x_k) \right] \cdot L^2_{n,k}(x)$$

$$D_{n,k}(x) \equiv (x - x_k) \cdot L^2_{n,k}(x)$$

- ## Hermite Table similar to DD Table

  - **Double** all entries & relabel $\{x_0 x_0 x_1 x_1 x_2 x_2\} \Rightarrow \{z_0 z_1 z_2 z_3 z_4 z_5\}$

  - Normal DD Table $\{z_0 z_1 z_2 z_3 z_4 z_5\}$

$$H^F_{2(2)+1}(x) = \sum_{k=0}^{5} f[z_0 z_1 \cdots z_k] \cdot (x - z_0) \cdot (x - z_1) \cdots (x - z_{k-1})$$

The Hermite polynomial allows us to take advantage of position and velocity data as a function of time. We could simply ignore the velocity data and apply Lagrange or Newtonian Divided Difference techniques to the time and position columns of the table to obtain an interpolating polynomial for x as a function of t. However, it is often the case that the velocity data is more accurate than the position data and so we would be foolish to ignore it; moreover there is no reason why we should not use **all** the available data. A table containing (n+1)-nodes, (n+1)-nodal values, and (n+1)-nodal derivatives allows us to match a polynomial with 2(n+1) independent conditions imposed on it. Thus for (n+1)-nodes, the degree of the resulting Hermite polynomial is 2(n+1)-1 = 2n+1. The formula for the Hermite polynomial of degree 2n+1 is the sum of terms of the form $f(x_k)*F_{n,k}(x) + f'(x_k)* D_{n,k}(x)$. The 2n+1 degree basis polynomials $F_{n,k}(x)$ and $D_{n,k}(x)$ are constructed from the square $[L_{n,k}(x)]^2$, the derivative $L_{n,k}'(x)$ of the Lagrange polynomials and $(x-x_k)$ so as to satisfy the (n+1) nodal constraints $H_{n,k}(x_k) = f(x_k)$ and (n+1) derivative constraints $H_{n,k}'(x_k) = f'(x_k)$.

Construction of the Hermite polynomial by this formula first requires construction of all n^{th} degree Lagrange polynomials and their derivatives, then formation of the basis polynomials $F_{n,k}(x)$ and $D_{n,k}(x)$ and multiplication by $f(x_k)$ and $f'(x_k)$ respectively, and finally, summation over all nodes k = 0,1,2, ..., n. Although straightforward, this procedure is computationally expensive. The next slide shows a Divided Difference Tableau that generates the Hermite polynomial in a much more efficient way.

3.8.1 Hermite Poly Example

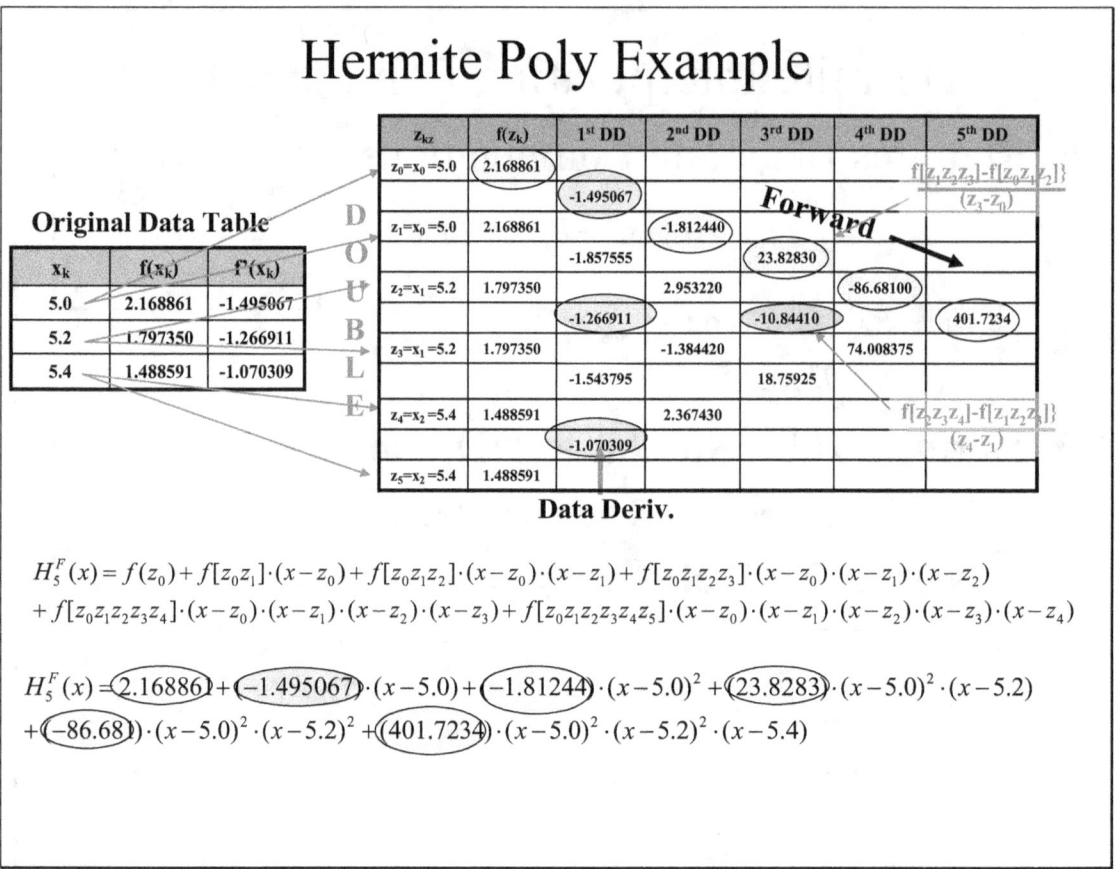

$$H_5^F(x) = f(z_0) + f[z_0 z_1] \cdot (x - z_0) + f[z_0 z_1 z_2] \cdot (x - z_0) \cdot (x - z_1) + f[z_0 z_1 z_2 z_3] \cdot (x - z_0) \cdot (x - z_1) \cdot (x - z_2)$$
$$+ f[z_0 z_1 z_2 z_3 z_4] \cdot (x - z_0) \cdot (x - z_1) \cdot (x - z_2) \cdot (x - z_3) + f[z_0 z_1 z_2 z_3 z_4 z_5] \cdot (x - z_0) \cdot (x - z_1) \cdot (x - z_2) \cdot (x - z_3) \cdot (x - z_4)$$

$$H_5^F(x) = 2.168861 + (-1.495067) \cdot (x - 5.0) + (-1.81244) \cdot (x - 5.0)^2 + (23.8283) \cdot (x - 5.0)^2 \cdot (x - 5.2)$$
$$+ (-86.681) \cdot (x - 5.0)^2 \cdot (x - 5.2)^2 + (401.7234) \cdot (x - 5.0)^2 \cdot (x - 5.2)^2 \cdot (x - 5.4)$$

The following Tableau computes the Hermite polynomial coefficients in a manner similar to the DD method. It is given without proof.

Starting from the original data table of nodal values and derivatives, we proceed by transferring the data to a DD table by writing down each node twice. In order to distinguish identical terms, we temporarily re-label them as follows: $z_0 = x_0$, $z_1 = x_0$, $z_2 = x_1$, $z_3 = x_1$, $z_4 = x_2$, $z_5 = x_2$; we place the corresponding functional values $f(x_k)$ in the second column of the table.

Proceeding as in a DD table we take the 1st DD for the points $z_0 = x_0$, $z_1 = x_0$ which yields an indeterminate value "0/0; we simply replace this indeterminate 1st DD value by the "known" derivative from the original data table corresponding to this node. The 1st DD for the next pair of points $z_1 = x_0$ and $z_3 = x_1$ yields a determinate value which is entered. The next pair $z_2 = x_1$, $z_3 = x_1$ again yields an indeterminate form and is replaced by the data derivative for the x_1-node. We continue down the column using the computed 1st DD when it is determinate and the data derivative when it is not.

Once the 1st column is in place we proceed as usual to complete the DD table out to the 5th DD as shown. Note that the original data table has only 3 nodes $\{x_0, x_1, x_2\}$ and yields a quadratic $P_2(x)$; however, the Hermite DD table derived from it has 6 nodes $\{z_0, z_1, z_2, z_3, z_4, z_5\}$ and yields a quintic $H_5(x)$. The coefficients for the forward (backward, central) Hermite polynomial are read directly off the table using the "z-notation" as:

$$H_5^F(x) = f[z_0] + f[z_0 z_1](x - z_0) + f[z_0 z_1 z_2](x - z_0)(x - z_1) + f[z_0 z_1 z_2 z_3](x - z_0)(x - z_1)(x - z_2)$$
$$+ f[z_0 z_1 z_2 z_3 z_4](x - z_0)(x - z_1)(x - z_2)(x - z_3)$$
$$+ f[z_0 z_1 z_2 z_3 z_4 z_5](x - z_0)(x - z_1)(x - z_2)(x - z_3)(x - z_4)$$

3.9 Cubic Splines

Cubic Splines

4 Pts., 3 Subsplines, 12 Unknowns

$$S(x) = \begin{cases} S_0(x) = a_0 + b_0(x-x_0) + c_0(x-x_0)^2 + d_0(x-x_0)^3 & x_0 \le x < x_1 \\ S_1(x) = a_1 + b_1(x-x_1) + c_1(x-x_1)^2 + d_1(x-x_1)^3 & x_1 \le x < x_2 \\ S_2(x) = a_2 + b_2(x-x_2) + c_2(x-x_2)^2 + d_2(x-x_2)^3 & x_2 \le x \le x_3 \end{cases}$$

4 Nodal Pts – 4 eqns

Total spline $S(x_k) = f(x_k)$

at nodes
$$S_0(x_0) = f(x_0) \; ; \; S_1(x_1) = f(x_1)$$
$$S_2(x_2) = f(x_2) \; ; \; S_2(x_3) = f(x_3)$$

Interior Pts – 6 eqns

Subsplines: match values,

1st and 2nd Deriv.

$x = x_1$

$x = x_2$

$$S_0(x_1) = S_1(x_1) \qquad S_1(x_2) = S_2(x_2)$$
$$S_0'(x_1) = S_1'(x_1) \qquad S_1'(x_2) = S_2'(x_2)$$
$$S_0''(x_1) = S_1''(x_1) \qquad S_1''(x_2) = S_2''(x_2)$$

End Pts – 2 eqns

Free(acceleration=0) $S_0''(x_0) = S_2''(x_3) = 0$

Clamped (end slope specified) $S_0'(x_0) = f'(x_0) \qquad S_2'(x_3) = f'(x_3)$

We can always fit a polynomial of degree n through a set of (n+1)- nodal points, but such a fit may not conform to the underlying function being sampled. Let's consider 4 points (n=3) and instead of simply fitting a cubic to the whole curve, break the fit up into regions I, II and III between successive pairs of points as shown in the figure. Now simply making linear fits between the 3 pairs of points will yield a jagged fit with the derivatives being undefined at the boundaries between regions I, II and II, III.   In order to obtain a smoother behavior we need to use a higher degree polynomial in each region and match their derivatives across the region boundaries. To match both 1st and 2nd derivatives we need a cubic fit in each region and we write the cubic sub-splines $S_0(x)$, $S_1(x)$, $S_2(x)$ in terms of fours sets of unknown coefficients $\{a_0,b_0,c_0,d_0\}$, $\{a_1,b_1,c_1,d_1\}$, $\{a_2,b_2,c_2,d_2\}$, respectively.

Thus we need to impose 12 conditions to solve for the 12 cubic spline coefficients. There are **4 nodal value equations** to satisfy $S_0(x_0)=f(x_0)$, $S_1(x_1)=f(x_1)$, $S_2(x_2)=f(x_2)$, $S_2(x_3)=f(x_3)$ and **3 pairs matching conditions at interior points** x_1, x_2: $S_0(x_1)=S_1(x_1)$, $S_1(x_2)=S_2(x_2)$; 1st derivatives: $S_0'(x_1)=S_1'(x_1)$, $S_1'(x_2)=S_2'(x_2)$; 2nd derivatives: $S_0''(x_1)=S_1''(x_1)$; $S_1''(x_2)=S_2''(x_2)$.

This leaves two additional conditions that need to be specified at the end points, so we may choose to set the 1st derivative (clamped) $S_0'(x_0)=f'(x_0)$; $S_2'(x_3)=f'(x_3)$ or 2nd derivative (free) $S_0''(x_0)=S_2''(x_3)=0$. Thus with a total of 12 conditions specified all the sub-spline coefficients can be determined. A matrix method for solving the resulting cubic spline equations are shown on the next slide.

3.9.1 Cubic Spline Algorithm

Cubic Spline Algorithm

1. Nodal values yield **"a"s** directly $\quad a_k = f(x_k) \; ; \; k = 0,1,2$

2. The **"c"s** satisfy one of the matrix eqns *Tri-Diagonal Matrix*

$$
\bullet \quad
\begin{bmatrix}
0 \\
\dfrac{3(a_2 - a_1)}{h_1} - \dfrac{3(a_1 - a_0)}{h_0} \\
\dfrac{3(a_3 - a_2)}{h_2} - \dfrac{3(a_2 - a_1)}{h_1} \\
0
\end{bmatrix}
=
\begin{bmatrix}
1 & 0 & 0 & 0 \\
h_0 & 2(h_0 + h_1) & h_1 & 0 \\
0 & h_1 & 2(h_1 + h_2) & h_2 \\
0 & 0 & 0 & 1
\end{bmatrix}
\cdot
\begin{bmatrix}
c_0 \\ c_1 \\ c_2 \\ c_3
\end{bmatrix}
\qquad \text{Free}
$$

Tridiagonal Matrix

$$
\bullet \quad
\begin{bmatrix}
\dfrac{3(a_1 - a_0)}{h_0} - 3 f'(x_0) \\
\dfrac{3(a_2 - a_1)}{h_1} - \dfrac{3(a_1 - a_0)}{h_0} \\
\dfrac{3(a_3 - a_2)}{h_2} - \dfrac{3(a_2 - a_1)}{h_1} \\
3 f'(x_3) - \dfrac{3(a_3 - a_2)}{h_2}
\end{bmatrix}
=
\begin{bmatrix}
2h_0 & h_0 & 0 & 0 \\
h_0 & 2(h_0 + h_1) & h_1 & 0 \\
0 & h_1 & 2(h_1 + h_2) & h_2 \\
0 & 0 & h_2 & 2h_2
\end{bmatrix}
\cdot
\begin{bmatrix}
c_0 \\ c_1 \\ c_2 \\ c_3
\end{bmatrix}
\qquad \text{Clamped}
$$

Tridiagonal Matrix

3. **"b"s** and **"d"s** are obtained from

$$
b_k = \frac{a_{k+1} - a_k}{h_k} - \frac{h_k}{3} \cdot (2c_k + c_{k+1}) \quad ; \quad d_k = \frac{c_{k+1} - c_k}{3 h_k}
$$

The cubic spline algorithm solves for the 12 coefficients which define the sub-splines. As shown in the slide, the particular order in which we solve for the "a"s, "b"s, "c"s, and "d"s is important; first the "a"s are set equal to the four nodal values; next, the "c"s are found from a solution to the matrix equation for either the "free" or "clamped" specification of end point values. Finally, the "b"s and then the "d"s are found from their indexed equations.

Adding more interior points follows the tri-diagonal matrix structure shown in this 4 point example; each new interior point adds a row with elements along the three diagonals and each has its index incremented by one. The vectors on the LHS of the equations are similarly extended by simply adding each interior row with incremented index.

Note that for the "free" case, c_1 and c_4 are both equal to zero and therefore the matrix equation in the n=4 free case reduces to a 2 x 2 matrix equation. Also note that in both the "free" and "clamped" cases the vector on the LHS may be written in a form that facilitates tabular computation using 1st divided differences, *viz.*,

$$3(a_1 - a_0)/h_0 = 3\,[f(x_1) - f(x_0)]/(x_1 - x_0) = 3f[x_0\,x_1] = 3\,\Delta a_0/h_0 \text{ , } etc. \text{ .}$$

3.9.2 Free Cubic Spline Example

Free Cubic Spline Example

-

	Nodal Table			Matrix Eqn	Recursions	
$h_k = \Delta x_k$	x_k	$a_k = f(x_k)$	Δa_k	c_k	b_k	d_k
0.1	0.1	-0.62049958	0.3365129	0	3.45508693	-8.9957933
0.1	0.2	-0.28398668	0.29058763	-2.698738	3.1852313	-0.94630333
0.1	0.3	+0.00660095	0.24182345	-2.982629	2.61607643	+9.9420966
	0.4	+0.24842440		0		

• Cubic Spline & Subsplines

$$S(x) = \begin{cases} S_0(x) = -.62049958 + 3.45508693(x-0.1) + (-8.9957933)(x-0.1)^3 & 0.1 \le x < 0.2 \\ S_1(x) = -0.28398668 + 3.18521313(x-0.2) + (-2.698738)(x-0.2)^2 + (-0.94630333)(x-0.2)^3 & 0.2 \le x < 0.3 \\ S_2(x) = 0.00660095 + 2.61707643(x-0.3) + (-2.982629)(x-0.3)^2 + 9.9420966(x-0.3)^3 & 0.3 \le x \le 0.4 \end{cases}$$

The table shows the computational set up for the 4 node "free" cubic spline which yields the coefficients for the 3 sub-splines given in the boxed equation for S(x). The table shows the nodal spacing h_k in the col#1, which are just the Δx_k for the original nodal table (cols#2,#3); col#4 gives Δa_k whose differences $\Delta(\Delta a_k)$ (not shown) appear in the matrix equation for the c_k. To summarize the table: the a_k are just the values in col#2 (nodal values), the c_k are obtained from the matrix equation, and the b_k and d_k are obtained from the recursion equations. Extension of this table to more points is straight forward. .

The cubic spline S(x) is given in the grey boxed equation; it consists of sub-splines defined for the three contiguous regions covering the full range of the spline.

3.10 *Tschebyshev Polynomials and Economization*

Tschebyshev Polynomials and Economization

- Defn "n-angle" trig fcn

$$T_n(\cos\theta) \equiv \cos n\theta \qquad 0 \le \theta \le \pi$$

$$\text{Zeros}: \ n\theta = (2k+1)\cdot\pi/2 \ ; \ k = 0,1,\cdots$$

- Defn in terms of $x = \cos\theta$

$$T_n(x) \equiv \cos(n\cdot\arccos x)$$

$$T_{n+1}(x) = 2xT_n(x) - T_{n-1}(x)$$

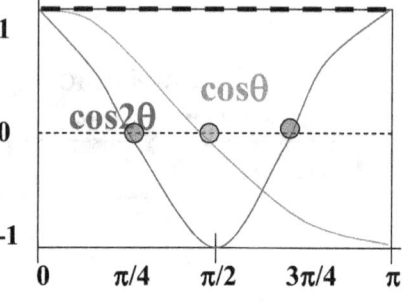

Tschebyshev	$\{1, x, x^2, \cdots\}$
$T_0(x) = 1$	$1 = T_0(x)$
$T_1(x) = x$	$x = T_1(x)$
$T_2(x) = 2x^2 - 1$	$x^2 = 2^{-1}(T_0 + T_2)$
$T_3(x) = 4x^3 - 3x$	$x^3 = 2^{-2}(T_3 + 3T_1)$
$T_4(x) = 8x^4 - 8x^2 + 1$	$x^4 = 2^{-3}(T_4 + 4T_2 + 3T_0)$
$T_5(x) = 16x^5 - 20x^3 + 5x$	$x^5 = 2^{-4}(T_5 + 5T_3 + 10T_1)$

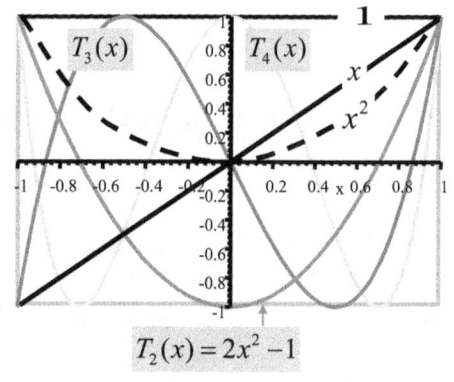

$$T_2(x) = 2x^2 - 1$$

The Tschebyshev polynomials are most naturally defined in terms of the "n-angle" trigonometric function as $T_n(\cos\theta) = \cos(n\theta)$ for $\theta \ \varepsilon \ [0, \pi]$ and has zeros at $\theta = 1\pi/2n, 3\pi/2n, 5\pi/2n,...$ as shown. The upper figure shows plots of the first three for n=0,1,2, *i.e.*, 1, $\cos\theta$, $\cos2\theta$.

If we set x= $\cos\theta$, then θ = arccos(x) and the polynomial is transformed into a function of the variable "x" yielding $T_n(x) = \cos(n \ \text{arccos}(x))$ where $\theta \ \varepsilon \ [0, \pi]$ maps to x ε [-1, +1]. The 2nd order recursion relation for $T_n(x)$ allows us to compute the next one from the previous two; the first two are defined to be $T_0(x) = 1$ and $T_1(x) = x$. Thus, the recursion yields $T_2(x) = 2x\{T_1(x)\} - T_0(x) = 2x*\{x\} - 1 = 2x^2-1$.

The table shows the first six Tschebyshev polynomials; the first two Tschebyshev polynomials $\{T_0(x), T_1(x)\}$ are identical to the first two monomials $\{1, x\}$. For n > 2, the Tschebyshev polynomials are linear combinations of only odd or only even monomials as shown in col#1. Col#2 gives the inverse by expressing the even (odd) monomials $\{1, x^1, x^2, ... \}$ in terms of the even (odd) Tschebyshev polynomials $\{T_0(x), T_1(x), T_2(x), T_3(x), T_4(x)\}$; it shows that each higher degree monomial is reduced in magnitude by a factor of two. Thus for n=2, x^2 is just the sum $T_0(x)+T_2(x)$ reduced by a factor of 2^{-1} while for n=3, x^3 is the sum $3T_1(x)+T_3(x)$ reduced by a factor of 2^{-2}. In general the Tschebyshev contributions to x^n decrease with monomial degree "n" by a factor $2^{-(n-1)}$. This reduction in magnitude together with the translation table provides the means for developing the so-called "Tschebyshev Economization" procedure described in the next slide.

The lower plot compares the first few Tschebyshev polynomials $\{T_0(x), T_1(x), T_2(x), T_3(x), T_4(x)\}$ with the monomials $\{1, x^1, x^2, ... \}$.

3.10.1   Tschebyshev Economization

Tschebyshev Economization

- **Taylor Series** $f(x) = \sin x = x - \dfrac{x^3}{6} + \dfrac{x^5}{120} + f^{(7)}(\xi) \cdot \dfrac{x^7}{7!}$ $x \in [0, \pi/2]$

- **Trunc. Error** $\varepsilon \leq \left| f^{(7)}(\xi) \cdot \dfrac{x^7}{7!} \right| = 1 \cdot \dfrac{\left(\frac{\pi}{2} - 0 \right)^7}{7!} = 4.7 \times 10^{-3}$

- **Rewrite from Tschebyshev Table**

$$P^{Tscheb}(x) = \underbrace{T_1}_{x} - \frac{1}{6} \cdot \underbrace{\left\{ 2^{-2}(T_3 + 3T_1) \right\}}_{x^3} + \frac{1}{120} \cdot \underbrace{\left\{ 2^{-4}(T_5 + 5T_3 + 10T_1) \right\}}_{x^5}$$

$$= T_1 \left(1 - \frac{3}{24} + \frac{10}{1920} \right) + T_3 \left(-\frac{1}{24} + \frac{5}{1920} \right) + \underbrace{\frac{1}{1920}}_{\substack{5.2 \times 10^{-4} \\ \mathbf{Drop}}} \underbrace{T_5}_{\leq 1}$$

- **"Economize"**

$$P^{Tscheb}(x) \cong \frac{1690}{1920} \underbrace{T_1}_{x} - \frac{75}{1920} \underbrace{T_3}_{4x^3 - 3x}$$

$$P^{Taylor}(x) \cong 1.000\overline{0}x - .16666x^3$$

$$\cong \frac{1915}{1920} \cdot x - \frac{300}{1920} \cdot x^3$$

$$\boxed{P^{Tscheb}(x) \cong .997396x - .15625x^3} \longleftarrow \text{—"Almost Taylor"}$$

Consider a Taylor polynomial $P_5(x)$ for the sin(x) over the interval x ε [0, $\pi/2$] ; since sin(x) is an odd function the error term is given by the next non-zero term which is x^7. The maximum error over this interval is found to be 4.7 x 10^{-3} ; The Tschebyshev Economization is a method that yields the same maximum error over the entire interval, but drops the Taylor x^5 term and changes the coefficients of the monomials {1, x^3} from their Taylor values {1, -1/6} to the Tschebyshev Economized values{.997396, -.15625}. This is made possible by the symmetric distribution of the Tschebyshev polynomials and the decreasing contributions of higher degree Tschebyshev polynomials by the factor $2^{-(n-1)}$.

To do this we simply use the table (last slide) to rewrite the monomials {1, x^3 , x^5} in terms of the $T_1(x)$, $T_3(x)$, $T_5(x)$, collect coefficients for each and recognize that the magnitude of the term (1/1920) $T_5(x)$ $\leq$ 1/1920 = 5.2 x 10^{-4} which is less than the Taylor error and so may be simply dropped. Using the table once again to translate $T_1(x)$, $T_3(x)$ back into monomial terms and collecting coefficients yields the pair {1915/1920, -300/1920}= {.997396, -.15625}.

3.11 *Bezier Parametric Curves & Computer Graphics*

Bezier Parametric Curves & Computer Graphics
- Hermite Polynomials $x(t), y(t)$ $t \, \varepsilon \, [0, 1]$

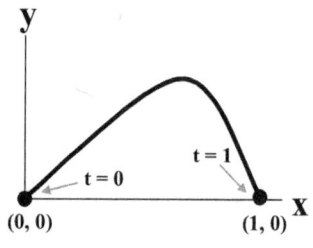

t_k	$x(t_k)$	1st DD	2nd DD	3rd DD
0	0			
		α_0		
0	0		$(1-\alpha_0)$	
		1		$(\alpha_1+\alpha_0-2)$
1	1		α_1-1	
		α_1		
1	1			

t_k	$x(t_k)$	$x'(t_k)$
0	0	α_0
1	1	α_1

General Parametric curves $x(t), y(t)$ that go through $(0,0)$ & $(1,0)$

t_k	$y(t_k)$	$y'(t_k)$
0	0	β_0
1	0	β_1

$$x(t) = 0 + \alpha_0 t + (1-\alpha_0)t^2 + (\alpha_0 + \alpha_1 - 2)t^2(t-1)$$

$$y(t) = \beta_0 t - \beta_0 t^2 + (\beta_0 + \beta_1)t^2(t-1)$$

An interesting application of Hermite polynomials to computer graphics is developed and illustrated in the next few slides. The basic idea is to express a curve or trajectory in the x-y plane parametrically in terms of a time variable t and to describe the x and y motions separately by the two cubic curves x(t) and y(t). The trajectory curve is constrained to begin and end on the x-axis; it starts at the origin (x,y)=(0,0) and ends a unit distance along the x axis at (x,y)=(1,0) just like a projectile under the influence of gravity. The parametric functions x(t) and y(t) (boxed equations at the bottom of the slide) have adjustable slope parameters (first derivatives) which change the shape of the resulting curve in the x-y plane within the x-interval [0,1].

Consider a set of data for times t_k, $x(t_k)$, $x'(t_k)$ displayed in the 1st table on the top left; the table shows two nodes t_0 =0 and t_1 =1 with nodal values of 0 and 1 and derivative values α_0 and α_1 respectively. We may double the entries and create the Hermite polynomial coefficients displayed in the DD table. This yields the cubic representation for x(t) given by the first boxed equation. The 2nd table on the lower left displays data for times t_k, $y(t_k)$, $y'(t_k)$ again with two nodes at t_0 =0 and t_1 =1 and new nodal values of 0 and 0 (y-value begins and ends at value 0) and derivative values β_0 and β_1. In a similar manner we create a DD table that yields a cubic representation for y(t) given by the second boxed equation.

This set of equations forms a parametric representation of the curve y(x) shown in the figure; we note that the origin in x,y coordinates (0,0) corresponds to parametric curve node t=0 and the point (1,0) corresponds to the parametric node t=1 for both the x(t) and y(t) cubic polynomials. Between the two nodes the values of x(t) and y(t) from the interpolating Hermite polynomials yields the rest of the curve. Clearly the curve always starts at the origin (0,0) and ends at (1,0) independent of how we choose the slope parameters α_0 , α_1 , β_0 , β_1. . These parameters will control the shape of the curve. The next slide details exactly how this is accomplished using "guide points" in the x-y plane.

3.11.1   Bezier Guide Points & Scale Factors

Bezier Guide Points & Scale Factors

- Compute tangents from left & right guide points

Left: $\dfrac{dy}{dx}(t=0) = \dfrac{\beta_0}{\alpha_0} \cdot \dfrac{s_0}{s_0} = \dfrac{y_L}{x_L}$

Right: $\dfrac{dy}{dx}(t=1) = \dfrac{\beta_1}{\alpha_1} \cdot \dfrac{s_1}{s_1} = \dfrac{y_R}{x_R - 1}$

Parametric curve as function of Guide Points

$$x(t) = \frac{x_L}{s_0}t + \left(1 - \frac{x_L}{s_0}\right)t^2 + \left(\frac{x_L}{s_0} + \frac{x_R - 1}{s_1} - 2\right)t^2(t-1)$$

$$y(t) = \frac{y_L}{s_0}t - \frac{y_L}{s_0}t^2 + \left(\frac{y_L}{s_0} + \frac{y_R}{s_1}\right)t^2(t-1)$$

y   Left Guide Pt: (x_L, y_L)

Slope = $(y_L\text{-}0)/(x_L\text{-}0)$

Right Guide Pt: (x_R, y_R)

Slope= $(y_R\text{-}0)/(x_R\text{-}1)$

$(0, 0)$ t = 0 $(1, 0)$ t = 1   **X**

y   Left Guide Pt: (x_L, y_L)

Right Guide Pt: (x_R, y_R)

$(0, 0)$ t = 0 $(1, 0)$ t = 1   **X**

In the x-y plane we may arbitrarily choose two guide points and with reference to the figure geometry we find the slope from the origin (0,0) to the "left guide point (red)" (x_L, y_L) is $dy/dx|_L = (y_L\text{-}0)/(x_L\text{-}0)$ and the slope from the right hand point (1,0) to the "right guide point (green)" (x_R, y_R) is $dy/dx|_R = (y_R\text{-}0)/(x_R\text{-}1)$. The parametric representation of x(t) and y(t) allows us to compute the two derivatives as $[(dy/dt)/(dx/dt)]_{t=0} = \beta_0/\alpha_0 = y_L/x_L$ and $[(dy/dt)/(dx/dt)]_{t=1} = \beta_1/\alpha_1 = y_R/(x_R\text{-}1)$. The problem is that we have 4 arbitrary slope parameters α_0, α_1, β_0, β_1 and only two equations; if we introduce arbitrary left and right guide point *scale parameters* s_0 and s_1, we may write $s_0\beta_0/(s_0\alpha_0) = y_L/x_L$ and $s_1\beta_1/(s_1\alpha_1) = y_R/(x_R\text{-}1)$. Equating numerators and denominators of each equation yields the four equations we need to solve for parameters α_0, α_1, β_0, β_1 at the expense of introducing the two new scale parameters s_0 and s_1. Upon substituting $\beta_0 = y_L/s_0$, $\alpha_0 = x_L/s_0$, $\beta_1 = y_R/s_1$, $\alpha_1 = (x_R\text{-}1)/s_1$ into the equations for x(t) and y(t) we arrive at the boxed set of Bezier curve equations.

The left and right guide point scale factors are seen to have the effect of "sliding" the guide point up or down the dotted slope lines; but in a graphics environment this is naturally accomplished by moving a mouse to arbitrarily select the slopes and actual distances of the left and right guide points to the two curve end points.

The following few slides illustrate a MatLab implementation (see Slides# 5-11 and 5-12) of these equations in which we fix the left guide point and then vary the right guide point. This illustrates the dynamics of moving a mouse on a computer screen so as to select the right guide point.

3.11.2   MatLab Bezier Curves

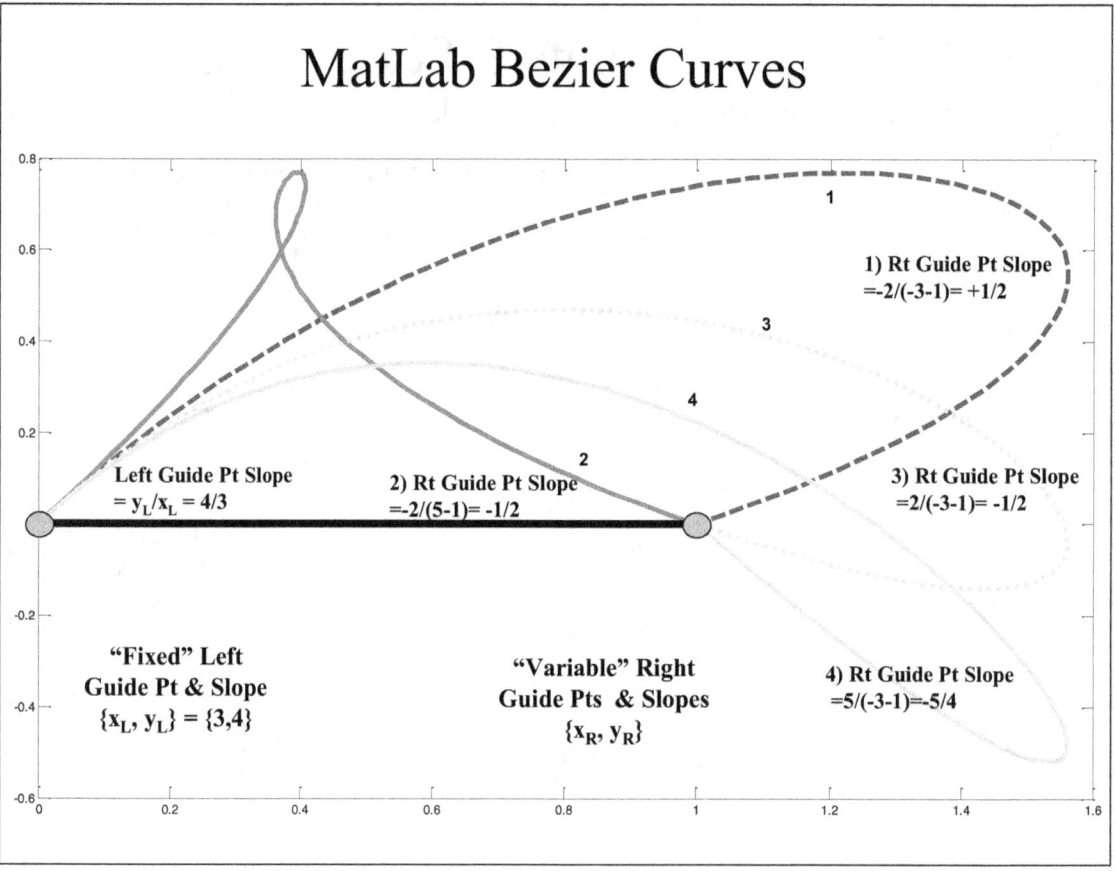

The following two slides illustrate a MatLab implementation of the Bezier equations for a *fixed left guide point* $(x_L, y_L) = (3,4)$ and *four different positions* for the right guide points, *viz.,*

(1) $(x_R, y_R) = (-3,-2)$, (2) $(x_R, y_R) = (+5,-2)$, (3) $(x_R, y_R) = (-3,+2)$, and (4) $(x_R, y_R) = (-3,+5)$.

The slopes at the right guide points are calculated on the slide from the formula slope $= y_R/(x_R-1)$ and correspond to the visual slopes in the plot. All curves are constrained to begin at $(0,0)$ and end at $(1,0)$ as indicated by the black line interval between the two large coordinate "dots." (See MatLab scripts on Slides# 5-11 and 5-12)

Observe that the red curve (2) $(x_R, y_R) = (+5,-2)$ and the green curve(3) $(x_R, y_R) = (-3,+2)$, have the same negative slope of -1/2, but the coordinate positions of their right guide points makes the curves radically different. Also note that the guide points are not shown because the x and y extent of this plot is too small; the next slide shows the guide points in a larger scale plot.

3.11.3   MatLab Bezier Curves-2

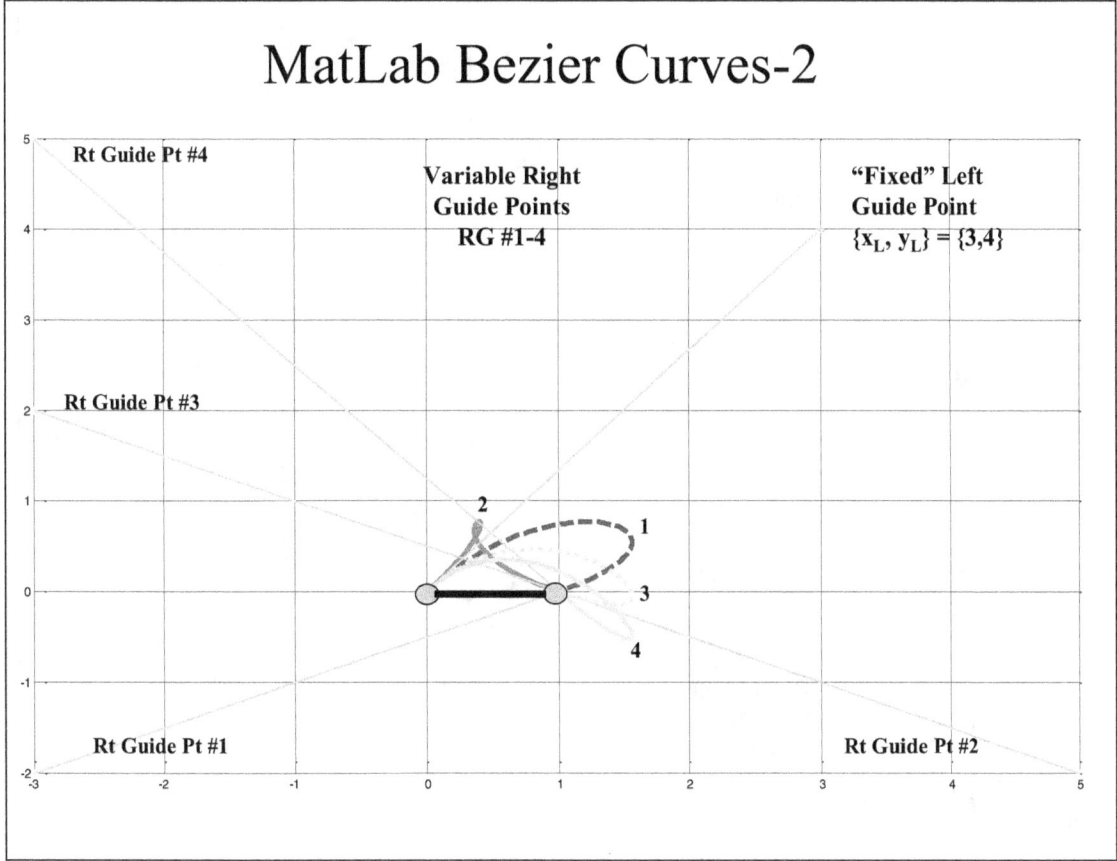

This slide corresponds to the parameters of the previous slide but the x and y values are now large enough to show the position of the right hand guide points at the expense of making the curves nearly invisible.   A description of the curves are repeated here for convenience: There is a *fixed left guide point* $(x_L, y_L) = (3,4)$ and *four different positions* for the right guide points, *viz.*,

(1) $(x_R, y_R) = (-3,-2)$, (2) $(x_R, y_R) = (+5,-2)$, (3) $(x_R, y_R) = (-3,+2)$, and (4) $(x_R, y_R) = (-3,+5)$.

The slopes at the right guide point are obtained from $y_R/(x_R-1)$ and correspond to the visual slopes in the plot. Observe, for example, that the red curve (2) and the dashed green curve(3) have equal negative slopes of $-1/2$, but the *diametrically opposed positions* of the right guide points for these two cases (which is now visible in this scaled plot) makes the curves radically different. (See MatLab scripts on Slides# 5-11 and 5-12)

3.11.4   MatLab Bezier Curves-3

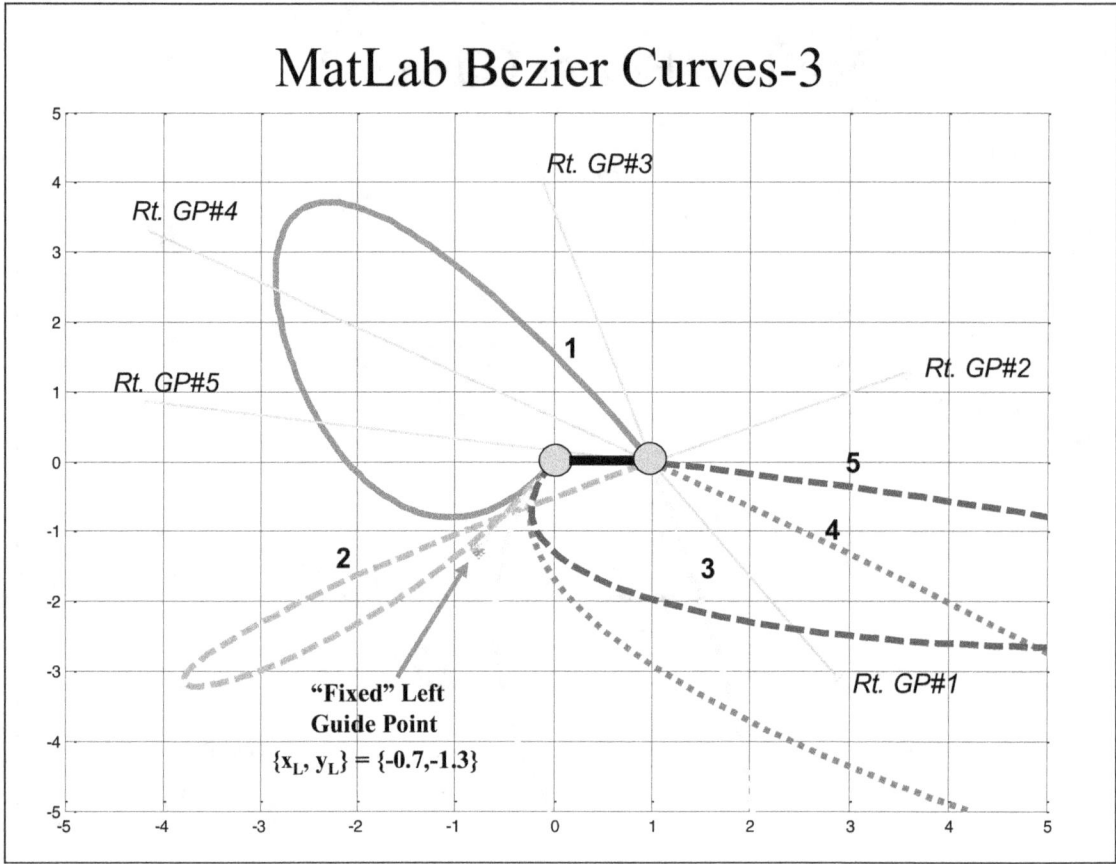

Here is another set of five Bezier curves for a fixed left guide point (red) located as shown at approximately $(x_L, y_L) = (-.7, -1.3)$ and five different positions for the right guide point (green) labeled Rt. GP#1 to Rt. GP#5 in the plot. All curves are constrained to begin at $(0, 0)$ and end at $(1, 0)$ as indicated by the two large coordinate "dots." This plot was made using the interactive MatLab script on Slide# 5-12 which accounts for the rather arbitrary location of the guide points; the MatLab script on Slide# 5-11 can be used instead to choose precise guide points and give more reproducible results. The large variation of the curves resulting from moving the right guide point (green) around is the reason why this is a good computer graphics tool; learning how to control the effects in a useful manner may take some practice.

4 Derivatives and Integrals

Derivatives and Integrals

- Numerical Derivatives
- Richardson Extrapolation
- Numerical Quadratures
- Improper Integrals
- Multiple Integrals

Interpolating polynomials over a set of nodal points yield numerical approximations to derivatives and integrals as well as an appropriate error bound. A number of derivative formulas are developed involving 2, 3, 4, or 5 nodal points and a general procedure for extrapolating results to higher truncation order gives a relationship between them. Numerical quadrature for a definite integral has a much richer legacy of methods that adapt to variation of the integrand, break it up into panels, or use different sets of interpolating polynomials with special properties that are efficient and especially useful when evaluating improper integrals.

It is important to note that differentiation is an unstable process involving a tradeoff between decreasing truncation and increasing round off errors as h decreases; on the other hand integration has a constant round-off error and decreasing truncation error with decreasing h and is thus stable.

4.1   *Computation using Polynomial Interpolates*

Computation using Polynomial Interpolates

- ## Differentiation
  - – Langrange & Taylor polynomials
  - – Richardson Extrapolation & Instability

- ## Integration - Numerical Quadrature
  - – Trapezoidal, Simpson
  - – Newton-Cotes (open, closed)
  - – Composite Trap_n , Simp_n
  - – Adaptive Quadrature
  - – Romberg = Trap_n + Richardson Extrapolation
  - – Gaussian Quadrature Methods
  - – Improper Integrals
    - Open Newton-Cotes
    - Transformation of Integration Limits
    - Orthonormal Quadratures
  - – Multiple integrals - Strips on a surface

$$I = \int_a^b f(x)dx = \underbrace{\sum_{k=0}^{n} a_k f(x_k)}_{\text{Wgtd Sum of Nodal Values}}$$

Interpolating polynomials over a set of nodal points yield numerical approximations to derivatives and integrals as well as an appropriate error bound. A number of derivative formulas are developed involving 2, 3, 4, or 5 nodal points and a general procedure for extrapolating results to higher truncation order gives a relationship between them.

A numerical quadrature is an explicit computation of the definite integral and is usually based on uniformly spaced nodal points over the integration interval [a,b] A set of methods using Lagrange polynomials fit to n+1 points defines the Newton-Cotes methods of degree n. The most well known are the n=1 trapezoidal and n=2 Simpson methods. Rather than fitting the nodal points to higher degree Lagrange polynomials, composite integration methods use trapezoidal or Simpson by breaking the interval [a,b] into panels and then summing. Other methods change the step size so as to adapt to the variation in the integrand and Gaussian quadrature departs from uniformly spaced nodal points using instead the zeros of the Legendre polynomials. Richardson extrapolation when combined with the composite trapezoidal method generates a very effective algorithm. Multiple integrals use a 2-dimensional nodal mesh and the techniques for single integrals are performed sequentially along x and then along the y coordinates of the "surface" nodal points.

4.1.1 Derivative of Lagrange Polynomial

Derivative of Lagrange Polynomial
1st Degree Lagrange Polynomial nodes x_0, x_1

$$f(x) = f(x_0) \cdot \frac{(x-x_1)}{(x_0-x_1)} + f(x_1) \cdot \frac{(x-x_0)}{x_1-x_0} + \frac{f''(\xi(x))}{2} \cdot (x-x_0) \cdot (x-x_1) \quad ; \quad \xi \in [x_0, x_0+h]$$

Differentiate wrt x

$$f'(x) = f(x_0) \cdot \frac{1}{(x_0-x_1)} + f(x_1) \cdot \frac{1}{(x_1-x_0)} + \frac{f''(\xi(x))}{2} \cdot [(x-x_1) + (x-x_0)]$$

$$+ \frac{f'''(\xi(x))}{2} \cdot \underbrace{\xi'(x)}_{\text{unkwn fcn}} \cdot \underbrace{(x-x_0) \cdot (x-x_1)}_{\text{vanishes at } x_0, x_1}$$

$x_{-1} = x_0 - h$ $x_1 = x_0 + h$

1st Deriv at node $x = x_0$

x_{-2} x_{-1} x_0 x_1 x_2

ξ

$$f'(x_0) = \frac{[f(x_0+h) - f(x_0)]}{h} - \frac{h}{2} \cdot f''(\xi(x)) \qquad \text{2 Pt}$$

A linear Lagrange interpolate and error term is written down for the nodes at x_0, x_1. Upon differentiation we obtain an expression for f'(x) which has the derivative of an unknown function $\xi'(x)$; however this term has factors of $(x-x_0)$ and $(x-x_1)$ which vanish if we evaluate the expression at one of the nodes x_0 or x_1. Thus, choosing $x=x_0$ we find the expression for the first derivative the f'(x_0) together with the error term. This expression is precisely the "calculus" approximation of Slide# 1-5, but now we have a truncation error estimate.

4.1.2 Derivative Formulas

Derivative Formulas
n^{th} Degree Lagrange Polynomial & Differentiate:

$$f(x) \equiv \underbrace{\sum_{k=0}^{n} L_{n,k}(x) \cdot f(x_k)}_{n-\text{pt Estimate}} + \underbrace{f^{(n+1)}(\xi(x)) \cdot \prod_{k=0}^{n} \frac{(x - x_k)}{(n+1)!}}_{\text{Truncation Error}} \quad ; \quad \xi \in [x_0, x_0 + h]$$

3-pt Derivative formulas

$$f'(x_0) = \frac{1}{2h} \cdot [-3f_0 + 4f_1 - f_2] + \frac{h^2}{3} \cdot f^{(3)}(\xi) \ ; \ \xi \in [x_0, x_2]$$

$$f'(x_0) = \frac{1}{2h} \cdot [f_1 - f_{-1}] - \frac{h^2}{6} \cdot f^{(3)}(\xi) \ ; \ \xi \in [x_{-1}, x_{+1}]$$

$x_{-1} = x_0 - h$ \qquad $x_1 = x_0 + h$

5-pt Derivative formulas

X_{-2} \quad X_{-1} \quad X_0 ξ X_1 \quad X_2

$$f'(x_0) = \frac{1}{12h} \cdot [f_{-2} - 8f_{-1} + 8f_1 - f_2] + \frac{h^4}{30} \cdot f^{(5)}(\xi) \ ; \ \xi \in [x_{-2}, x_{+2}]$$

$$f'(x_0) = \frac{1}{12h} \cdot [-25f_0 + 48f_1 - 36f_2 + 16f_3 - 3f_4] - \frac{h^4}{5} \cdot f^{(5)}(\xi) \ ; \ \xi \in [x_0, x_4]$$

We can generalize the 2-pt derivative formula by differentiating an nth degree Lagrange polynomial to obtain an n+1 point differentiation formula with its error term. Again we evaluate the expression at the nodes in order to eliminate the term involving the unknown derivative $\xi'(x)$.

Several 3-point and 5-point derivative formulas are given using a the notation $\{x_{-2}, x_{-1}, x_0, x_1, x_2\}$ for the nodes and $\{f_{-2}, f_{-1}, f_0, f_1, f_2\}$ for the corresponding nodal values. Note that the nodes are given by $x_k = x_0 + k\,h$ for all positive and negative integers "k" and further we use the convenient notation

$$f_0 = f(x_0), \quad f_{-1} = f(x_0 - h), \quad f_1 = f(x_0 + h), \text{ or in general } f_k = f(x_0 + k\,h).$$

This notation allows the formulas to be easily adapted to various positions in the nodal table by a simple translation of the x coordinates to the desired position. For example, suppose we need a 3-pt formula for the derivative at the node x_{-2}, i.e., $f'(x_{-2})$ instead of $f'(x_0)$. To do this we simply shift everything to the left by 2 nodal steps by making the following transformations

$$x_0 \rightarrow x_0 - 2h = x_{-2};$$
$$x_1 \rightarrow x_1 - 2h = (x_0 + h) - 2h = x_{-1};$$
$$x_2 \rightarrow x_2 - 2h = (x_0 + 2h) - 2h = x_0.$$

These substitutions into the first 3-pt derivative formula yields the "new" 3-pt derivative formula (with the same error term) using the three left-most values $\{x_{-2}, x_{-1}, x_0\}$

$$f'(x_{-2}) = [-3f_{-2} + 4f_{-1} - f_0]/(2h)$$

Similarly shifting by 1 nodal step $x_0 \rightarrow x_0 - 1h$ yields another equivalent formula for the first derivative at $f'(x_{-1})$ using the three central values $\{x_{-1}, x_0, x_1\}$

$$f'(x_{-1}) = (1/2h)[-3f_{-1} + 4f_0 - f_1]/(2(-h))$$

Finally, to evaluate the derivative at the right end point x_2, we re-label it as "x_0" and step backwards h → -h to x_{-1} and x_{-2} to find $f'(x_0) = [-3f(x_0) + 4f(x_0-h) - f(x_0-2h)]/(2(-h)) = [-3f_0 + 4f_{-1} - f_{-2}]/(2(-h))$ (Note the "-h".)

4.1.3 Derivative Formulas Application

Derivative Formulas Application

Label	x_n	$f(x_n)$	$f'(x_n)$
x_0	-0.3	-0.20431	0.35785
x_1	-0.1	-0.08993	0.78595
x_2	+0.1	+0.11007	1.2141
x_3	+0.3	+0.39569	1.6422

(1) $\quad f'(x_0) = \dfrac{1}{2h} \cdot [-3f_0 + 4f_1 - f_2] + \dfrac{h^2}{3} \cdot f^{(3)}(\xi) \; ; \; \xi \in [x_0, x_2]$

(2) $\quad f'(x_0) = \dfrac{1}{2h} \cdot [f_1 - f_{-1}] - \dfrac{h^2}{6} \cdot f^{(3)}(\xi) \; ; \; \xi \in [x_{-1}, x_{+1}]$

$x_0 = -0.3$ Use (1) with $x_0 \; x_1 \; x_2$ $f'(-0.3) = \dfrac{1}{2(.2)} \cdot [-3(-0.20431) + 4(-0.08993) - (+0.11007)]$

$x_1 = -0.1$ Use (1) with x_1, x_2, x_3 $f'(-0.1) = \dfrac{1}{2(.2)} \cdot [-3(-0.08993) + 4(+0.11007) - (+0.39569)]$

Better use (2) re-label: $x_0 \; \boxed{x_1} \; x_2 \; x_3$

       as $x_{-1} \; \boxed{x_0} \; x_{+1}$ $f'(x_0) = \dfrac{1}{2(.2)} \cdot [(+0.11007) - (-0.20431)]$

$x_2 = +0.1$ Better use (2) re-label: $x_0 \; x_1 \; \boxed{x_2} \; x_3$

       as $x_{-1} \; \boxed{x_0} \; x_{+1}$ $f'(x_0) = \dfrac{1}{2(.2)} \cdot [(+.39569) - (-0.08993)]$

or Use (1′)   re-label: $x_0 \; x_1 \; \boxed{x_2} \; x_3$

"backward" steps   as $x_{-2} \; x_{-1} \; \boxed{x_0}$ $f'(+0.1) = \dfrac{1}{2(-.2)} \cdot [-3(+0.11007) + 4(-0.08993) - (-.20431)]$

$h \to -h$

$x_3 = +0.3$ Use (1) with "backward" steps $h \to -h$ to yield

Use (1′) re-label: $x_0 \; x_1 \; x_2 \; \boxed{x_3}$ (1′) $f'(x_0) = \dfrac{1}{2(-h)} \cdot [-3f_0 + 4f_{-1} - f_{-2}]$

       as $x_{-2} \; x_{-1} \; \boxed{x_0}$ $f'(+0.3) = \dfrac{1}{2(-.2)} \cdot [-3(+0.39569) + 4(+0.11007) - (-0.08993)]$

The first derivative formulas of the last slide are explicitly applied to the table of functional values to yield the first derivative column. The calculations are shown for each point using translated forms of forward step formula (1) and (2) and the "backward step " formula (1′) ; the latter formula results from letting $h \to -h$ in (1) and noting that

$\qquad\qquad f_1 = f(x_0 + h) \to f(x_0 - h) = f_{-1}$;similarly, $f_2 = f(x_0 + 2h) \to f(x_0 - 2h) = f_{-2}$.

The "best" formula to use is clearly the one whose points "straddle" the desired evaluation point, thus allowing us to use the symmetric formula (2); when this is not possible, the closest points are next best.

For x_0 there are no straddling points so we must use Eq.(1) with points x_0, x_1, x_2

For x_1 we have the choice of again using Eq.(1) but now with points x_1, x_2, x_3, or instead using the symmetric formula of Eq.(2) with points x_0, x_1, x_2 ,straddling x_1

For x_2 we have the choice of again using Eq.(1) but now with points x_2, x_3, x_4, or instead using the symmetric formula of Eq.(2) with points x_1, x_2, x_3 ,straddling x_2

For x_3 there are no straddling points so we must use Eq.(1′) with backward steps starting at x_3 using the points x_3, x_2, x_1 replacing respectively x_0, x_{-1}, x_{-2} in the formula.

The explicit calculations are shown and the "best" results for the derivative are displayed in the third column of the table.

4.1.4 Differentiation – Taylor Tricks

Differentiation – Taylor Tricks

- 3rd Degree Taylor Polynomials abt x_0

$$f(x_0 + h) = f(x_0) + f'(x_0)h + f''(x_0)\frac{h^2}{2!} + f'''(\xi(x))\frac{h^3}{3!} \quad ; \quad \xi \in [x_0, x_0 + h]$$

$$f(x_0 - h) = f(x_0) + f'(x_0)(-h) + f''(x_0)\frac{(-h)^2}{2!} + f'''(\xi(x))\frac{(-h)^3}{3!}$$

- Subtract to cancel even terms 1st Deriv

$$f'(x_0) = \frac{[f(x_0 + h) - f(x_0 - h)]}{2h} - f^{(3)}(\xi(x)) \cdot \frac{h^2}{6}$$ **3 Pt**

- Add to cancel odd terms ($\rightarrow$4th Degree) 2nd Deriv

$$f''(x_0) = \frac{1}{h^2} \cdot [f(x_0 - h) - 2f(x_0) + f(x_0 + h)] - \frac{h^2}{12} \cdot f^{(4)}(\xi(x))$$ **3 Pt**

We can use polynomials other than Lagrange to develop derivative formulas; this is accomplished by fitting the polynomial to the table of nodal points, then differentiating and evaluating at a nodal point.

If we write the Taylor polynomials for $f(x_0+h)$ and $f(x_0-h)$ and subtract them we obtain the 3-point formula for the first derivative we had on the previous slide; adding them eliminates the first derivative $f'(x_0)$ and yields a new 3-point formula for the second derivative $f''(x_0)$

As we shall see all the first derivative formulas can be related *via* the process of Richardson Extrapolation which we shall cover next. In this regard, we note that the structure in "h" for the three-point derivative is determined by continuing to higher orders (not truncating with a derivative evaluated at $\xi(x)$) and we find the infinite series structure jumps by two orders in h, *viz.*,

$f'(x_0) = [f(x_0+h) - f(x_0-h)]/(2h) - f^{(3)}(x_0)h^2/3! + f^{(5)}(x_0)h^4/5! - f^{(7)}(x_0)h^6/7!+...$ (3-pt deriv. jumps by h^2)

Compare this with the two-point derivative of Slide# 4-3 which only jumps by one order in h

$f'(x_0) = [f(x_0+h) - f(x_0)]/h - f^{(2)}(x_0)h^1/2! + f^{(3)}(x_0)h^2/3! - f^{(4)}(x_0)h^3/4!+...$ (2-pt deriv. jumps by h^1)

Note that for the second derivative $f''(x_0)$ we need to take the two expansions out one order further with error term $f^{(iv)}(\xi(x))$ because the terms containing $f'''(\xi(x))$ cancel upon addition leaving no error term.

4.2 *Instability of Numerical Derivatives*

Instability of Numerical Derivatives

- $\varepsilon_{RO} \sim h^{-1}$ versus
- $\varepsilon_{Trunc} \sim h^1$ or $h^2 \cdots$
- Desire h small to reduce ε_{Trunc}
- But small h gives large ε_{RO}

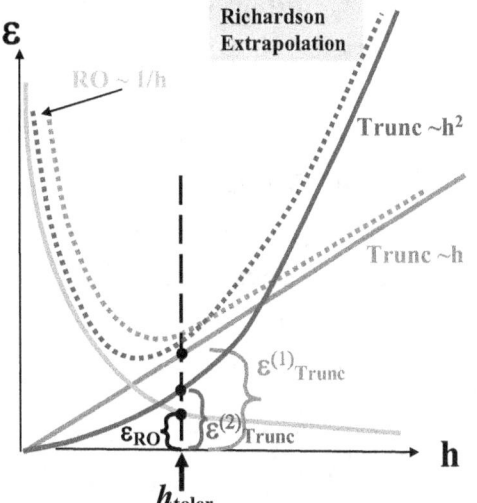

- 3-pt formula has less trunc error for fixed h
  - Fix h at sufficiently large value to yield tolerable ε_{RO}
  - Extrapolate to higher order method (**Blue**) reduce ε_{Trunc} for same h_{toler}
  - → Richardson Extrapolation

When computing a numerical derivative, round off error always varies inversely with stepsize h as $\varepsilon_{RO} \sim h^{-1}$, while truncation error varies directly with some power of h as $\varepsilon_{trunc} \sim h, h^2, h^3, \ldots$. Clearly, if we reduce the stepsize h in order to reduce the truncation error, this also has the effect of increasing the round off error. Thus, numerical differentiation is an unstable computation with respect to stepsize because we must choose a stepsize h that trades off these two competing error sources.

One trade off approach is to fix the stepsize at a value that gives a tolerable round off error h_{toler} (vertical dashed line in figure) and then decrease the (larger) truncation error by going to a higher order in h as illustrated by the linear and quadratic truncation error curves in the figure. The plot shows the round off error ε_{RO} (green) drawn as the hyperbola h^{-1} and the two truncation error curves linear (red) $\varepsilon_{trunc} \sim h$ and quadratic (blue) $\varepsilon_{trunc} \sim h^2$; the total errors for each is also shown by the dashed curves. The relative magnitudes of the round off error and linear and quadratic truncation errors is illustrated in the figure by the curve intersections with the vertical dashed line $h = h_{toler}$. It is seen that the truncation errors are both larger than the round off error

Clearly, so long as the truncation error dominates *i.e.*, $\varepsilon_{trunc} > \varepsilon_{RO}$, it makes sense to continue reducing the truncation error by going to higher order in h; for fixed $h = h_{toler}$ the round off error is unchanged and this procedure just beats down the larger truncation error.

Richardson Extrapolation is a technique that generates a new functional approximation for a numerical derivative (or integral) by using its known structure in "h" to eliminate lower orders of h thereby going to successively higher order truncation errors $\varepsilon_{trunc} \sim h, h^2, h^3, \ldots$. This allows us to "squeeze out" as much truncation error as possible for a fixed round off error ε_{RO}. We develop Richardson Extrapolation on the next slide.

4.3 *Richardson Extrapolation*

Richardson Extrapolation

1. Let $N_1[h]$ denote the algorithm (deriv., integral, *etc.*) computed using stepsize h

Trunc. Error Decreases $\longrightarrow$			
RO Error $N_1[h]$			
Increases $N_1[h/2]$	$N_2[h]$		
$N_1[h/4]$	$N_2[h/2]$	$N_3[h]$	
$N_1[h/8]$	$N_2[h/4]$	$N_3[h/2]$	$N_4[h]$

ε_{RO}^{max}

2. Halve stepsize down 1st Col & compute $N_1[h/2]$ *etc.*, until RO error ε_{RO}^{max} is maximum ***tolerable*** magnitude

3. Then extrapolate each row to right to reduce the truncation error

Interpretation: Combine *poorer data* (stepsize h) with *better data* (stepsize $h/2$) in order to reduce truncation error

$$N_k[h] = N_{k-1}[h/2] + \frac{N_{k-1}[h/2] - N_{k-1}[h]}{(2^{jump})^{k-1} - 1} \; ; \; k = 2, 3, \cdots$$

Example: Quadratic h-structure: $M = N_1[h] + a_1 h^2 + a_2 h^4 + a_3 h^6 + \cdots$ *jump=2*

Row#2 $N_2[h] = N_1[h/2] + \dfrac{N_1[h/2] - N_1[h]}{3}$

Row#3
$$\begin{cases} N_2[h/2] = N_1[h/4] + \dfrac{N_1[h/4] - N_1[h/2]}{3} \\[2mm] N_3[h] = N_2[h/2] + \dfrac{N_2[h/2] - N_2[h]}{15} \end{cases}$$

Richardson Extrapolation takes advantage of the known h-dependence of truncation error as we change the approximation order from linear to quadratic to cubic, *etc.*. The table illustrates the basic extrapolation structure with the truncation error decreasing across the rows from left to right and the round off error increasing down each column as we halve the stepsize. The extrapolation formula given in the boxed equation is for a quadratic from term-to-term (jump=2), *i.e.*, $M = N_1[h] + a_1 h^2 + a_2 h^4 + a_3 h^6 + a_4 h^8 + \cdots$; for the naïve derivative calculation (Slides# 4-11, 4-12) we have linear from term-to-term (jump=1), *i.e.*, $M = N_1[h] + a_1 h^1 + a_2 h^2 + a_3 h^3 + a_4 h^4 + \cdots$. Note that both of these depend upon the detailed h-structure of the error terms and moreover this extrapolation procedure can be generalized to different stepsize changes (other than halving) so long as they have a patterned structure in stepsize "h".

The first entry in the table $N_1[h]$ represents a derivative computation with stepsize h (e.g., "2-point" derivative formula). The next several entries in the first column display "2-point" derivative results obtained by halving stepsizes for each entry $N_1[h/2]$, $N_1[h/4]$, $N_1[h/8]$. Using pairs of first column entries we can construct "new" derivative formulas down the second column $N_2[h]$, $N_2[h/2]$, $N_2[h/4]$ and similarly down the next two columns we use pairs to construct $N_3[h]$, $N_3[h/2]$ and $N_4[h]$ completing the table. The methods for constructing these new derivative formulas are given explicitly below the table.

The basic idea is to specify an initial stepsize h that has negligible round-off error and then reduce the relatively large truncation error by halving stepsize in each row. We start by using two stepsizes h and h/2 and compute $N_1[h]$ and $N_1[h/2]$; the truncation error for $N_1[h/2]$ is improved by factor of 2 and although the round-off error is now twice its original small value, there is a net improvement. Next, instead of ignoring the *poorer derivative estimate* $N_1[h]$, we use it together with $N_1[h/2]$ to generate an *improved new formula* "$N_2[h]$" in col#2 which has 1/4 the truncation error of $N_1[h]$ and leaves the round-off error at its row#2 value.

Adding the next row#3 starts by computing $N_1[h/4]$ and then pairing it with $N_1[h/2]$ to generate a second member of col#2 $N_2[h/2]$. Row#3 is completed by pairing "$N_2[h/2]$" and "$N_2[h/4]$" to generate col#3 "$N_3[h]$" which has $\sim h^3$ truncation error and maintains the small row#3 round-off error. Eventually, we reach an optimal point where the *decrease in truncation error* equals the *increase in round-off error* for the given row and we stop.

4.3.1 Mathematical Details

Mathematical Details

Infinite Taylor expansion gives "exact" derivative:

$$M \equiv f'(x_0) = \underbrace{\frac{[f(x_0+h)-f(x_0-h)]}{2h}}_{\equiv N_1[h]} \underbrace{- f^{(3)}(x_0)\cdot\frac{h^2}{3!}}_{\equiv -a_1 h^2} \underbrace{- f^{(5)}(x_0)\cdot\frac{h^4}{5!}}_{\equiv -a_2 h^4} \underbrace{- f^{(7)}(x_0)\cdot\frac{h^6}{7!}}_{\equiv -a_3 h^6} - \cdots \quad \text{[Note: } jump = 2\text{]}$$

h-dependence

$$M = N_1[h] + a_1 h^2 + a_2 h^4 + a_3 h^6 + \cdots \qquad (1)$$

Rewrite with
$h \rightarrow h/2$

$$M = N_1[h/2] + a_1 (h/2)^2 + a_2 (h/2)^4 + a_3 (h/2)^6 + \cdots \quad (2)$$

Eliminate a_1:
$4*(2)-(1)$

$$M = \underbrace{\frac{4N_1[h/2]-N_1[h]}{3}}_{\equiv N_2[h]} + 0 + \tilde{a}_2(h)^4 + \tilde{a}_3(h)^6 + \cdots \quad (3)$$

Repeat process with $N_2[h]$

$$M = N_2[h] + 0 + \tilde{a}_2(h)^4 + \tilde{a}_3(h)^6 + \cdots \quad (3')$$

$$M = N_2[h/2] + 0 + \tilde{a}_2(h/2)^4 + \tilde{a}_3(h/2)^6 + \cdots \quad (4)$$

$$M = \underbrace{\frac{16N_2[h/2]-N_2[h]}{15}}_{\equiv N_3[h]} + 0 + \tilde{\tilde{a}}_3(h/2)^6 + \cdots \quad (5)$$

For the formal derivation we start with the Taylor-derived 3-point derivative formula $N_1[h]$ shown and note that the Taylor series yields the following truncation error structure as a function of stepsize h

$$M = N_1[h] + a_1 h^2 + a_2 h^4 + a_3 h^6 + ..., \quad (jump=2)$$

where, the "a" s are the Taylor coefficients of the order of h^2, h^4, h^6, ... and M denotes the exact value of the derivative expressed as a convergent infinite sum. The same sum for M can be written by letting $h \rightarrow h/2$

$$M = N_1[h/2] + a_1 (h/2)^2 + a_2 (h/2)^4 + a_3 (h/2)^6 + ...,$$

Upon eliminating the h^2 term we find a new expression for M which starts with a new (higher order) derivative computation algorithm $N_2[h]$ and a leading error term $\sim h^4$. Continuing the process with $N_2[h]$ and $N_2[h/2]$ yields a new $N_3[h]$ with leading error term $\sim h^6$.

Note that the structure for the derivative might not always change by h^2 as in the Taylor expansion above so the extrapolation procedure will differ from the one described above. Moreover, we may not always halve the interval we might instead let $h \rightarrow h/3$ or it might be different from row-to-row, $e.g.$ $h \rightarrow h/2$, $h \rightarrow h/3$, $h \rightarrow h/4$.

As a simple example, if the h-dependence of the algorithm $N_1[h]$ changes by h^1 ($jump=1$), we write

$$M = N_1[h] + a_1 h^1 + a_2 h^2 + a_3 h^3 + \; ; \quad M = N_1[h/2] + a_1 (h/2)^1 + a_2 (h/2)^2 + a_3 (h/2)^3 +$$

and we would find instead

$$N_2[h] = 2*N_1[h/2] - N_1[h] = N_1[h/2] + \{N_1[h/2] - N_1[h]\} \quad (jump=1)$$

This is taken care of by the "jump" parameter in the general formula of Slide# 4-8.

4.3.2 Extrapolate from 3-point to 5-point Derivative Formula

Extrapolate from 3-point to 5-point Derivative Formula

- **3-pt derivative formula** $f'(x_0) = N_1[h] = \dfrac{[f(x_0+h) - f(x_0-h)]}{2h}$

$$M = N_1[h] + a_1 h^2 + a_2 h^4 + a_3 h^6 + \cdots$$

$$N_1[2h] = \frac{[f(x_0+2h) - f(x_0-2h)]}{2(2h)}$$

- **Extrapolate**

$$N_2[h] = \frac{4N_1[h] - N_1[2h]}{3} = \frac{4\dfrac{[f(x_0+h)-f(x_0-h)]}{2h} - \dfrac{[f(x_0+2h)-f(x_0-2h)]}{2(2h)}}{3}$$

$$N_2[h] = \frac{1}{12h} \cdot [8f(x_0+h) - 8f(x_0-h) - f(x_0+2h) + f(x_0-2h)]$$

- $\rightarrow$ | **5-pt** $\quad f'_0 = \dfrac{1}{12h}[f_{-2} - 8f_{-1} + 8f_1 - f_2]$ |

$$\varepsilon_{trunc} \leq \frac{h^4}{30} \left| f^{(5)}(\xi) \right|$$

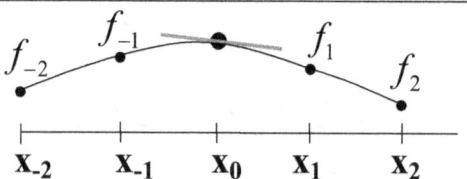

Extrapolation procedures are quite general and can be applied to any numerical algorithm whose analytic dependence upon the stepsize h is known. It can also be applied theoretically to generate 5-point derivative methods from the Taylor 3-point method as the example of this slide illustrates.

We recall that the Taylor 3-point derivative is given by $N_1[h]$ shown in the slide; if we look at a poorer estimate using h→2h given by $N_1[2h]$, then we can align these in a column as row#1= $N_1[2h]$ and row#2=$N_1[h]$ so the stepsize is halved and we can immediately write down $N_2[h]$= {4 $N_1[h]$- $N_1[2h]$}/ 3 . Upon substituting the actual expressions for the Taylor 3-point derivative for these two stepsizes we obtain the 5-point formula shown. If we "carry" the Taylor truncation error terms in the calculation we also arrive at the correct 5-point truncation error shown. Note that we use the simpler algebraic formula here instead of the more numerically efficient general formula given in Slide# 4-8, *viz.*,

$$N_2[h] = (4N_1[h] - N_1[2h])/3 = N_1[h] + (N_1[h] - N_1[2h])/3$$

The inset figure shows the 5-point nature of the derivative computation involving points symmetrically located about the central point x_0. Thus although x_0 is not involved in the functional evaluations needed to obtain the derivative, it is the 5th point at which the derivative is estimated.

4.4 *Extrapolation Table Stopping Criteria*

Extrapolation Table Stopping Criteria

- Extrapolating the *naive* 5 sd derivative table (see Slide#1-6) $f(x) = e^x$

$f'_{exact}(x_0 = 1) = 2.71828$

$$\varepsilon_{tot} \le \left| f'(x_0) - \hat{f}'[x_0, h] \right| \le \frac{2e_{5sd}}{h_k} + \frac{Mh_k}{2} \ ; \ h_k = 2^{-k}$$

- Recall Slide#1-27

- At k=6: $\varepsilon_{RO} \sim 2e_{5sd}/h_6 = 2(10^{-5})/.015625 = .00128$ $\underbrace{\varepsilon_{RO}}$ $\underbrace{\varepsilon_{trunc}}$

     $\varepsilon_{trunc} \sim \max[f^{(3)}]*h_6/2 = e^1 *.015625/2 = .0212$

- Note table below for $k = 7$ $h_7 = .0078$ $\rightarrow$ $h_7^2 = 6.1 \times 10^{-5}$

- Thus the 2nd order truncation term $O(h^2)$ is a correction that is at the limits of the 5 sd *machine precision, viz.,* $h_7^2 = 6.1 \times 10^{-5}$

- Thus stop at k=7 since *additional rows* cannot improve trunc error and will only serve to increase the RO error

- RO error dominates the lower rows k = 8, 9, 10 and no amount of extrapolation will improve the results!   **Trunc. Error Decreases →**

	5sd		O(h)	O(h²)	O(h³)	O(h⁴)	O(h⁵)
RO	**k**	**h_k**	**N1[h]**	**N2[h]**	**N3[h]**	**N4[h]**	**N5[h]**
Error	6	0.01562500	2.7328	0	0	0	0
incr.	→7	0.00781250	2.7264	2.7200	0	0	0
↓	8	0.00390620	2.7136	2.7008	2.6944	0	0
	9	0.00195310	2.7648	2.8160	2.8544	2.8773	0
	10	0.00097656	2.7648	2.7648	2.7477	2.7325	2.7228

The simple 2-point algorithm we applied to find the derivative of e^x at x=1 is partially reproduced in the table starting at k=6 corresponding to a stepsize $h_6 = 2^{-6} = .015625$. The resulting 2-point derivative values are listed in the column headed by $N_1[h]$ with truncation error O(h). Note that here we have linear from term-to-term h-structure, *i.e.*, $M = N_1[h] + a_1h^1 + a_2h^2 + a_3h^3 + a_4h^4 + \cdots$ so we use the boxed formula on Slide# 4-8 with jump=1 .In computing the derivative (see Slides# 1-26,1-27) we subtracted two functional evaluations which for 5 significant digits leads to a round off error $e_{5sd} \sim 10^{-5}$ for each; thus a bound on the RO error is obtained by adding the individual errors and dividing by stepsize h , *viz.*, $\varepsilon_{RO}(N_1[h_6]) \sim 2 *e_{5sd} / h_6 = 2 *10^{-5} /.015625 = .00128$. The linear truncation error for this derivative computation is $\varepsilon_{trunc}(N_1[h_6]) \sim \max[f''(x)] h_6 /2 = e^{(1)} * .015625 / 2 = .0212$. (See MatLab script on Slide# 5-10)

Since the truncation error, .0212, is 16 times the round off error, .00128, it makes sense to go to the next row $h_7 = 2^{-7}$ where we double the small round off error to $\varepsilon_{RO}(N_1[h_7]) = 2(.00128) = .00256$ and simultaneously halve the dominant truncation error to $\varepsilon_{trunc}(N_1[h_7]) = .0212/2 = .0106$.

Note that the truncation error is still dominant by a factor of 4, but now we have two values in the N_1 column and we can use Richardson Extrapolation to expand the table to the right to find $N_2[h_7]=2.7200$ with reduced truncation error $\sim O(h^2)$. The question is whether a third row (k=8) will allow us to extrapolate further to $O(h^3)$ by going down one more row and the across two columns; going down another row doubles RO to .00513 and halves truncation error in the N_1 column to .0053 so they cancel out one another. However, the truncation error in the next N_2 column is $\varepsilon_{trunc}(N_2[h_7]) \sim \max[f'''(x)] (h_7)^2 /3! = e^{(1)} * (.0078)^2 /3! = 2.7 \ 10^{-5}$, which is much smaller than the .00513 RO error and is also at the precision limits of a 5 sd machine. Thus, we would stop at k=7 and just perform the one extrapolation to the right. Note that the best extrapolated result for 5 sd is for k=7 in the $N_2[h]$ column, *viz.*, 2.7200 .

4.4.1 Extrapolation Tables 5sd & 8sd

Extrapolation Tables 5sd & 8sd

e = 2.718281828

5sd

		O(h)	O(h²)	O(h³)	O(h⁴)	O(h⁵)
k	h_k	N1[h]	N2[h]	N3[h]	N4[h]	N5[h]
6	0.01562500	2.7328	0	0	0	0
→7	0.00781250	2.7264	2.7200	0	0	0
8	0.00390620	2.7136	2.7008	2.6944	0	0
9	0.00195310	2.7648	2.8160	2.8544	2.8773	0
10	0.00097656	2.7648	2.7648	2.7477	2.7325	2.7228

8sd

		O(h)	O(h²)	O(h³)	O(h⁴)	O(h⁵)
k	h_k	N1[h]	N2[h]	N3[h]	N4[h]	N5[h]
8	0.00390625	2.7235584	0	0	0	0
→9	0.001953125	2.7209216	2.7182848	0	0	0
10	0.000976563	2.7197440	2.7185664	2.7186603	0	0
11	0.000488281	2.7191296	2.7185152	2.7184981	2.7184750	0
12	0.000244141	2.7181057	2.7170818	2.7166040	2.7163334	2.7161906

Increasing our computational precision from 5 to 8 sd shows that we can make further improvements by going to k=9, h_9=.0019531, but again adding another row doubles the round off error and subsequent extrapolation along the new row only reduces the total error to the extent that the truncation error is larger than the round off error. With each new row the round off error doubles so that eventually round off error dominates and extrapolation along that row has negligible effect on the total error. The stopping point argument is similar to that for 5 sd. (See MatLab script on Slide# 5-10)

For 8 sd we have at k=8

$$\varepsilon_{RO}(N_1[h_8]) \sim 2\, e_{8sd} / h_8 = 2*10^{-8} /.003906 = 5.12\ 10^{-6}$$

On the other hand the linear truncation error for this computation is

$$\varepsilon_{trunc}(N_1[h_9]) \sim max[f''(x)]\, h_{10} /2 = e^{(1)} *.003906 / 2 =.0053$$

At k=9 we halve the stepsize to $h_9 = 2^{-9} = .003906 / 2$ the round off error doubles while the truncation halves, but is still much larger than the round-off error, *viz.*,

$$\varepsilon_{RO}(N_1[h_9]) \sim 2e_{8sd} / h_9 = 2*10^{-8} /.00195 = 10^{-5} ;$$
$$\varepsilon_{trunc}(N_1[h_9]) = e^{(1)} *.00195 / 2 =.0027$$

Richardson Extrapolation in this case yields a result correct to 7 sd

$$N_1[h_9]+(N_1[h_9]- N_1[h_8]) = 2.7209216 +(2.7209216-2.7235584) = \mathbf{2.7182848}$$

Again we can determine if another row in the extrapolation table will make further improvements by calculating the new truncation error $O(h^3)$ for that row and compare it to twice the round off error $2*10^{-5}$. We find

$$\varepsilon_{trunc}(N_3[h_{10}]) \sim max[f''(x)]\, (h_{10})^3 /4! = e^{(1)} * (9.8\ 10^{-4})^3 /4! = 10^{-10.}$$

This is much smaller than the round off error and is also beyond the precision limits of an 8 sd machine, so we would stop at k=9 and just one extrapolation to the right. Note that the best extrapolated result for 8 sd is for k=9 in the $N_2[h]$ column, *viz.*, **2.71828**48 .which has the first 6 digits correct.

4.5 *Integral of Lagrange Polynomial*

Integral of Lagrange Polynomial

- Quadrature

$$I = \int_a^b f(x)dx \cong \sum_{k=0}^{n} a_k f(x_k)$$

<u>Wgtd Sum of Nodal Values</u>

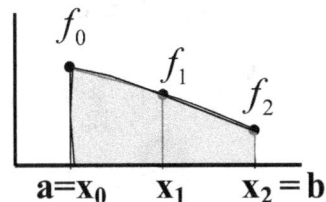

- Substitute Lagrange Interpolation Polynomials

$$I = \int_a^b f(x)dx \equiv \int_a^b \sum_{k=0}^{n} L_{n,k}(x) \cdot f(x_k)dx + \int_a^b R_n^L(x;\xi)dx$$

$$I = \sum_{k=0}^{n} a_k f(x_k) + \frac{1}{(n+1)!} \int_a^b \left[\prod_{k=0}^{n}(x-x_k) \right] f^{(n+1)}(\xi(x))dx$$

Quadrature · *Truncation Error*

$$a_k \equiv \int_a^b L_{n,k}(x)dx$$

One method to obtain a numerical quadrature is to break up the integration interval [a,b] into a uniform linear grid of n+1 equally spaced nodal points and to expand f(x) as an nth degree Lagrange polynomial fit to the n+1 points. Substituting the Lagrange polynomial (with error term) into the integral yields a weighted sum of the nodal values with coefficients a_k determined by the easily computed integrals of the Lagrange polynomials over the interval [a,b]. An upper bound to the truncation error is found by first maximizing the $(n+1)^{st}$ derivative for $\xi \ \varepsilon$ [a,b] and then integrating the remaining product polynomial in x over the interval. The resulting set of quadratures defines the Newton-Cotes methods of degree n.

4.5.1 Integration Formulas (Quadrature)

Integration Formulas (Quadrature)

- **Trapezoidal** (linear: 2 pt) n=1

$$I_T = \frac{h}{2}[f_0 + f_1] - \frac{h^3}{12} f''(\xi)$$

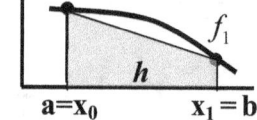

- **Simpson** (quadratic: 3 pt) n=2

$$I_S = \frac{h}{3}[f_0 + 4f_1 + f_2] - \frac{h^5}{90} f^{(4)}(\xi)$$

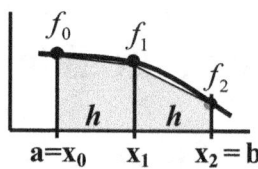

- **Newton-Cotes** (degree n: (n+1)-pts)
  - *Closed* $a = x_0 \; ; b = x_n$
  - *Open* $a \neq x_0 \; ; b \neq x_n$

  Used for Improper integrals
  with singularities at end pts

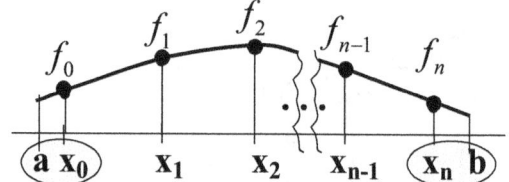

The most well known Newton-Cotes methods are the n=1 trapezoidal and n=2 Simpson methods which result from linear (nodes x_0, x_1) and quadratic (nodes x_0, x_1 , x_2) Lagrange polynomials respectively as illustrated in the two upper figures. The general Newton-Cotes method of degree n is illustrated in the bottom figure but the corresponding integration quadrature formulas are not given.

Because the n=2 Simpson method integrates using a quadratic polynomial we expect that the truncation error $\sim h^3$ will integrate to give an error $\sim h^4$ rather than $\sim h^5$ shown in the Simpson equation. This reason for this is that the integration of the cubic polynomial multiplying $f^{(3)}(\xi)$ vanishes and hence we must use the next higher degree term (quartic polynomial) *times* $f^{(4)}(\xi)$ in the Lagrange expansion which yields the $\sim h^5$ truncation error shown.

There are two types of quadratures, namely "closed Newton-Cotes" which has nodes at the integration interval end points, i.e., x_0=a and x_n=b and "open Newton-Cotes" which has the first and last node not at the endpoints. The main reason for using the latter open version is to allow a quadrature for improper integrals whose integrand diverges at the end points or for which one or both of the end points is infinity.

The higher degree Newton-Cotes methods fit large degree Lagrange polynomials to the interval [a,b] taken as a whole and this does not allow adaption to a function that may change rapidly in some regions and slowly in others. Generally speaking, it is more advantageous to break the integral over the interval [a,b] into a number of trapezoidal and Simpson "panels" thereby keeping the fitting method simple and allowing for adaption to functional variations by decreasing the stepsize for panels in regions of rapid functional variation and conversely increasing the stepsize in regions of smoother functional variation. These composite methods are described on the next slide.

4.5.2 Composite Methods

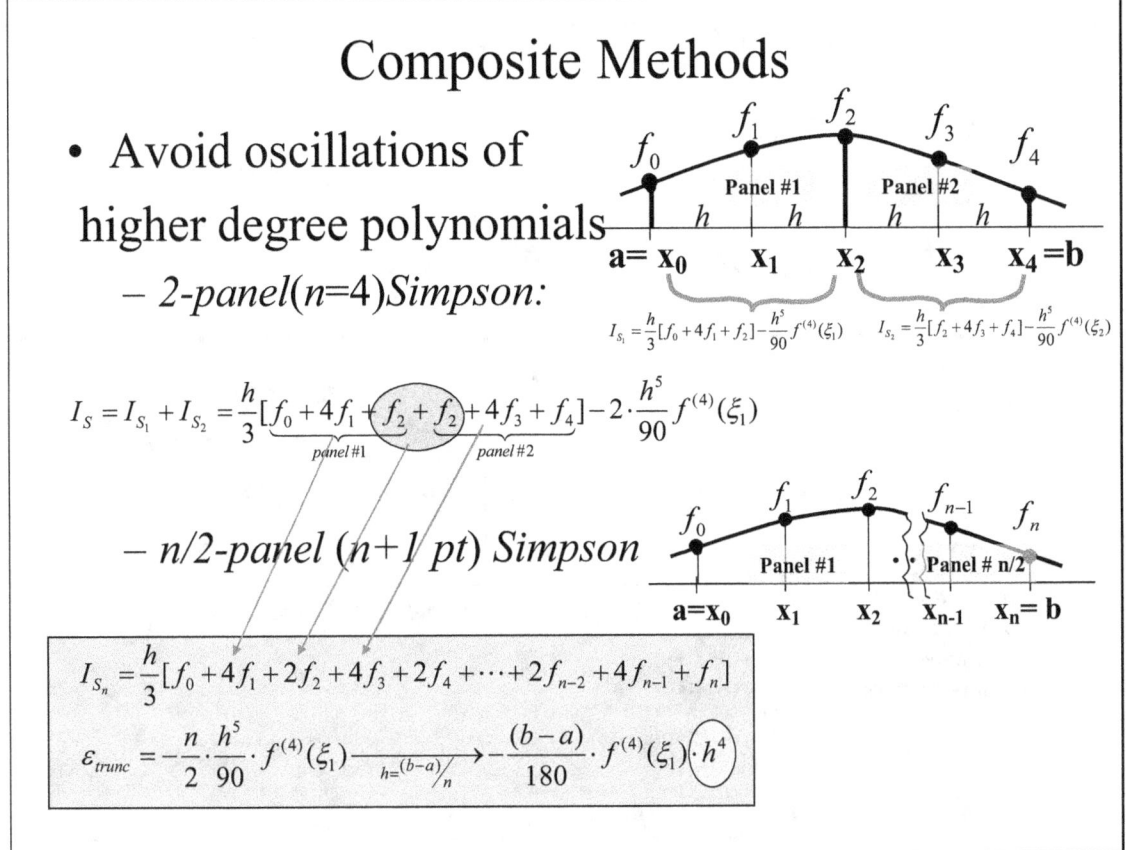

Composite Methods

- **Avoid oscillations of higher degree polynomials**
  - *2-panel($n=4$)Simpson:*

$$I_{S_1} = \frac{h}{3}[f_0 + 4f_1 + f_2] - \frac{h^5}{90} f^{(4)}(\xi_1) \qquad I_{S_2} = \frac{h}{3}[f_2 + 4f_3 + f_4] - \frac{h^5}{90} f^{(4)}(\xi_2)$$

$$I_S = I_{S_1} + I_{S_2} = \frac{h}{3}[\underbrace{f_0 + 4f_1 + (f_2}_{panel \#1} + \underbrace{f_2) + 4f_3 + f_4}_{panel \#2}] - 2 \cdot \frac{h^5}{90} f^{(4)}(\xi_1)$$

  - *$n/2$-panel ($n+1$ pt) Simpson*

$$I_{S_n} = \frac{h}{3}[f_0 + 4f_1 + 2f_2 + 4f_3 + 2f_4 + \cdots + 2f_{n-2} + 4f_{n-1} + f_n]$$

$$\varepsilon_{trunc} = -\frac{n}{2} \cdot \frac{h^5}{90} \cdot f^{(4)}(\xi_1) \xrightarrow{h=(b-a)/n} -\frac{(b-a)}{180} \cdot f^{(4)}(\xi_1) \cdot h^4$$

For the reasons just discussed, we consider Simpson and trapezoidal composite methods. The composite Simpson method is shown for a set of 5 points $\{x_0, x_1, x_2, x_3, x_4\}$ which are broken up into Panel#1 and Panel#2; the composite Simpson method simply applies the Simpson formula with error term to each panel and sums them. When we join the two panels each with Simpson coefficient pattern $\{1\ 4\ 1\} + \{1\ 4\ 1\}$ we obtain the composite coefficient pattern $\{1\ 4\ (1+1)\ 4\ 1\}$ where "middle term has a 2 resulting from a pair of common panel edges. Each additional panel has a common edge so for example for three panels $\{1\ 4\ 1\} + \{1\ 4\ 1\} + \{1\ 4\ 1\}$ we obtain $\{1\ 4\ 2\ 4\ 2\ 4\ 1\}$ where the "2"s represent the common edge between adjacent panels. This is generalized to give the boxed formula of the composite Simpson method across the interval [a,b].

Each Simpson panel will contribute a different amount to the error because the 4th derivative $f^{(4)}(\xi)$ for each panel is different; however choosing the maximum value of this derivative over the entire interval [a,b], we can bound the error by multiplying the *maximum single panel error* by the number of panels n/2 as shown in the boxed equation .

Note that n must be even in order to fit Simpson panels across the interval [a,b]; the stepsize is defined as $h = (b-a)/n$. Finally, note that even though the truncation error appears to vary with stepsize as $\sim h^5$, the multiplication by the number of panels: $\varepsilon_{trunc} \sim h^5 * n = h^5 * (b-a)/h = (b-a)\ h^4$ reduces it to h^4 dependence

4.5.3 Composite Trapezoid & Simpson

Composite Trapezoid & Simpson

Trapezoid n-panel (n+1 pt)

$$I_{T_n} = \frac{h}{2}[f_0 + 2f_1 + \cdots + 2f_{n-1} + f_n]$$

$$\varepsilon_{trunc} = -n \cdot \frac{h^3}{12} \cdot f^{(2)}(\xi) \xrightarrow[h=(b-a)/n]{} -\frac{(b-a)}{12} \cdot f^{(2)}(\xi) \cdot h^2$$

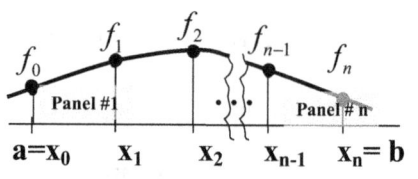

Simpson n/2-panel (n+1 pt)

$$I_{S_n} = \frac{h}{3}[f_0 + 4f_1 + 2f_2 + 4f_3 + 2f_4 + \cdots + 2f_{n-2} + 4f_{n-1} + f_n]$$

$$\varepsilon_{trunc} = -\frac{n}{2} \cdot \frac{h^5}{90} \cdot f^{(4)}(\xi_1) \xrightarrow[h=(b-a)/n]{} -\frac{(b-a)}{180} \cdot f^{(4)}(\xi_1) \cdot h^4$$

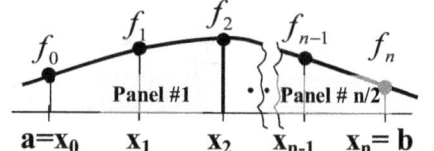

MatLab for n-panel
Composite Simpson

I_n(a,b,np,eq)

$x \in [a,b]$; np = #panels
eq = input function
e.g., eq = 'x.^4'
(suitable for lists ".* ")

See Slide#5-21

```
function In = I_n(a,b,np,eq) %input fcn for lists with "." e.g., eq = 'x.^4'
f=inline(eq,'x');
h1=(b-a)/(2*np);
x_1=a+[1:2:2*np]*h1;
if np == 1
    In= (h1/3)*(f(a)+f(b)+4*f((a+b)/2));   % One Simpson Panel
elseif np >= 2
x_2=a+[2:2:2*np-1]*h1;
In= (h1/3)*(f(a)+f(b)+4*sum(f(x_1))+2*sum(f(x_2) )); end
```

The derivation of the trapezoidal n-panel method is similar to that for Simpson, only simpler. (We show both here for convenient comparison.) Because each panel has only two evaluation points, every point except the first and last is common to two panels and thus has a coefficient of "2". The coefficient pattern for 5 points $\{x_0, x_1, x_2, x_3, x_4\}$ results from the sum of n=4 trapezoidal panels, *viz.*, {1 1}, {1 1}, {1 1}, {1 1} which yields the composite coefficients {1 2 2 2 1}. For n+1 points, this generalizes to the boxed equation which has n- trapezoidal panels with stepsize h=(b-a)/n. A suitable n-panel algorithm is shown in the lower box which is easily implemented on any programmable calcularor. The sum is simplified by using two lists (odd and even coordinates), then taking the sums

I = (h/3)[f(a) + f(b) + 4*sum f(x_1)+2*sum f(x_2)]

Note that the function *eq* must be expressed in terms of ".* "; ".^ " type operations as shown.

Just as in the Simpson case the truncation error is bound by the product of the number of trapezoidal panels n with the largest single panel error; it yields a truncation error $\varepsilon_{trunc} \sim h^2$

Although, the composite trapezoidal method is only a linear fit for each panel, it does have the advantage that the number or panels n can be even or odd and thus adding panels or changing their stepsize is easily done. Moreover, we shall find that when combined with an extrapolation technique which decreases the truncation error, it becomes a very powerful and versatile alternative known as Romberg integration.

4.6 *Stability of Composite Quadrature Methods*

Stability of Composite Quadrature Methods

- **Truncation Error:** down one order $\boxed{\varepsilon_{trunc}^{(n)} \approx const. \cdot n \cdot h^n = (b-a) \cdot h^{n-1}}$

- **RO Error:** independent of stepsize h

  RO at each point $f(x_k) = \hat{f}(x_k) + e(x_k)$ $\boxed{\varepsilon_{RO}^{n/2} = \frac{n}{2} \cdot \varepsilon_{RO}^{Panel} \le n \cdot h \cdot \varepsilon^{max} = (b-a) \cdot \varepsilon^{max}}$

  One panel $\varepsilon_{RO}^{Panel} = \frac{h}{3} \cdot [e_0 + 4e_1 + e_2] \le \frac{h}{3} \cdot 6 \cdot \underbrace{\max\{e_0, e_1, e_2\}}_{\equiv \varepsilon}$

- **Implications:** .

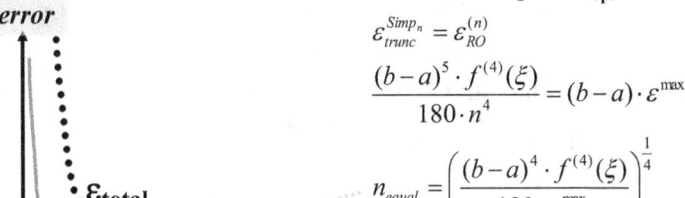

1. Can decrease stepsize h
   (increase # panels)
   w/o increasing RO error

2. Diminishing returns from
   stepsize reduction (**h decr**)
   since constant RO error
   is the limiting factor

3. **VERY STABLE!!**

Crossover pt & n_{opt}

$$\varepsilon_{trunc}^{Simp_n} = \varepsilon_{RO}^{(n)}$$

$$\frac{(b-a)^5 \cdot f^{(4)}(\xi)}{180 \cdot n^4} = (b-a) \cdot \varepsilon^{max}$$

$$n_{equal} = \left(\frac{(b-a)^4 \cdot f^{(4)}(\xi)}{180 \cdot \varepsilon^{max}} \right)^{\frac{1}{4}}$$

Recall the stability issues we have seen with numerical derivatives that arose because of the opposite behavior of the round-off and truncation error as a function of stepsize h.
It turns out that the composite integration methods are extremely stable with respect to round-off and in fact the stepsize can be decreased as much as needed without any round off instability occurring.
For Newton-Cotes numerical integration the truncation error for a degree n polynomial is $\varepsilon_{trunc} \sim h^n * h = h^{n+1}$ because we multiply by the stepsize to obtain the integral (Note that n=2 Simpson is an exception to this rule and actually is one order better $\varepsilon_{trunc} \sim h^4$.)
For a composite method we add n/2 panels and the truncation error drops down one order because n*h = (b-a) absorbs one factor of h as the constant interval b-a; thus the composite Simpson truncation is down one order $\varepsilon_{trunc} \sim h^4$.
Now the round off error is computed by adding up the RO errors for each functional evaluation over all the composite panels. As shown in the slide computation the RO error in n/2 Simpson panels is bounded by taking the max error over the whole interval [a.b], say, e_m and computing $\varepsilon_{RO} <$ (h/3)*(e_m+4e_m+e_m)*(n/2)= (n*h)*e_m=(b-a)e_m= *constant*
The figure shows a horizontal green line representing the constant round-off error e_{RO} and the composite Simpson (red) truncation error $\varepsilon_{trunc} \sim h^4 \sim n^{-4}$ plotted versus the number of points n=(b-a)/h. Clearly, we can reduce the stepsize (increase n) with impunity since the stepsize does not affect the fixed round off error. However, there comes a point of diminishing returns when the truncation error curve crosses the green RO curve and the two errors are equal. Decreasing the stepsize much beyond this is not useful since the increasingly smaller reductions in truncation error are dominated by the larger round off error

4.7 *Adaptive Quadrature*

Adaptive Quadrature

- One panel simpson yields large error
- Adapt nodal sampling rate to functional variation
- Need test procedure to monitor functional variation

  subject to some tolerance ε_{TOL}
- Use known dependence upon "h"

1-panel
$$I_S[h] = \frac{h}{3}[f_0 + 4f_1 + f_2];$$

2-panel
$$I_S[h/2] = \frac{h}{3}[f_0 + 4f_1 + 2f_2 + 4f_3 + f_4];$$

f(x)

Panel #1   Panel #2

h/2   h/2   h/2   h/2

a   (a+b)/2   b

in 1 & 2 Panel to estimate trunc. error for 2-panel result (rather than improve it)

$$I = I_S[h] + \varepsilon[h]; \qquad \varepsilon[h] = -\frac{h^5}{90} f^{(4)}(\xi_1)$$

$$I = I_S[h/2] + \varepsilon[h/2]; \quad \varepsilon[h/2] = -2 \cdot \frac{(h/2)^5}{90} f^{(4)}(\xi_2) = -\frac{2}{32} \frac{h^5}{90} f^{(4)}(\xi_2) \cong \frac{1}{16} \varepsilon[h]$$

Subtract and solve for ε [h/2]

$$0 = I_S[h] - I_S[h/2] + \underbrace{\varepsilon[h] - \varepsilon[h/2]}_{=16\varepsilon[h/2]} \Rightarrow$$

- **TEST:** ε [h/2] < ε_{TOL} ?

$$\boxed{\varepsilon[h/2] = \frac{1}{15} \cdot |I_S[h] - I_S[h/2]|}$$

Adaptive quadrature is not a single method, but rather an approach that changes the stepsize so as to adapt to the variation in the integrand. Clearly, if we look at the f(x) function plotted in the figure, a single Simpson panel with stepsize h will not capture the variation on the right hand side; although halving the stepsize and computing a composite of two panels will improve things, the right hand panel will obviously need to be split again.

The key to developing such an algorithm is to find a technique that determines when the function starts to change rapidly without plotting it. One such technique is to compute the integral twice, once as a single panel and a second time by halving the stepsize and computing the composite two panel result. Instead of using these two results to do an extrapolation we use them to estimate the two-panel truncation error ε[h/2] as shown in the boxed equation and then test this value against the desired truncation error tolerance ε_{TOL} . If the test is satisfied we are done and we can accept the result (and why not improve using Richardson extrapolation since we can); otherwise we halve the intervals again and test each half against $\varepsilon_{TOL}/2$. We continue in this manner splitting and testing against half the previous tolerance until all parts of the integral satisfy the test. A flow chart describing this process is shown on the next slide.

Derivatives and Integrals

4.7.1 Adaptive Quadrature Algorithm Flow

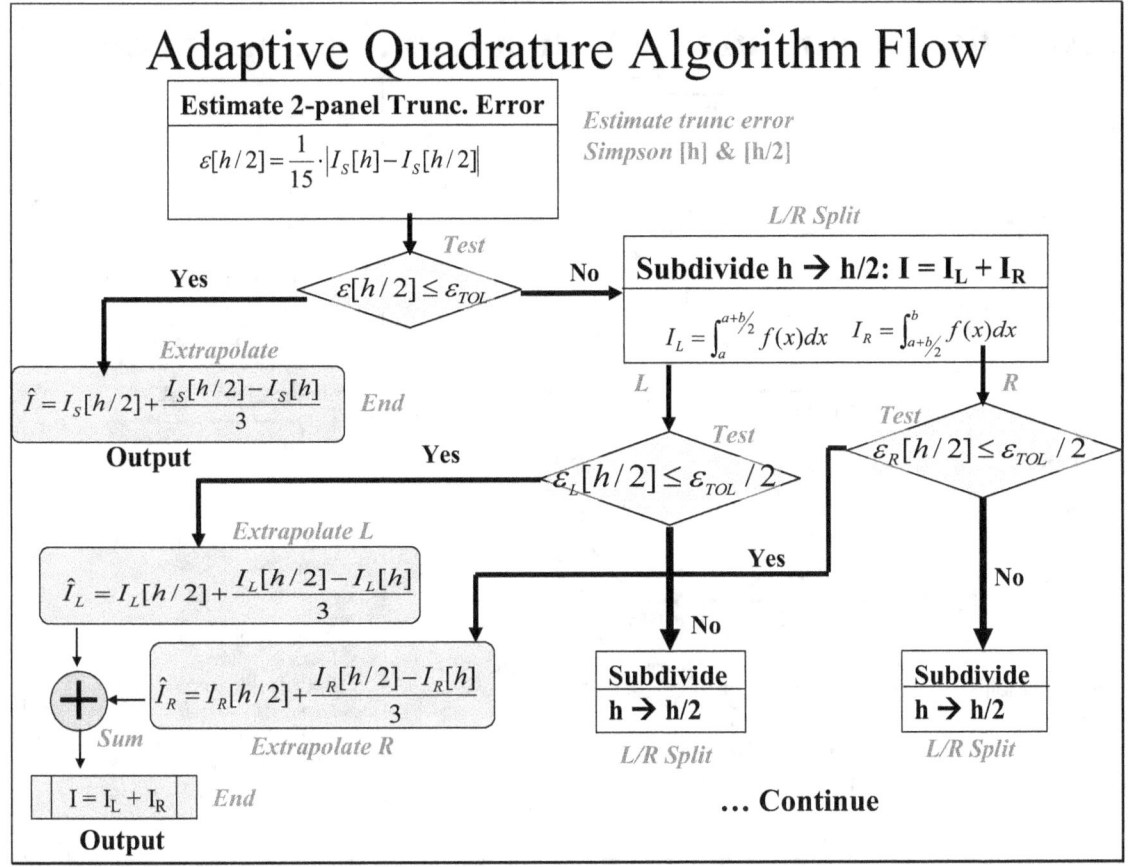

This flow diagram visualizes the algorithm described on the previous slide.

Step#1) Estimate the 2-panel truncation error by performing 2 quadratures.

Step#2) Test 2-panel truncation errors against their allocated tolerances. First time through there is one composite panel with allocation $\varepsilon[h/2] < \varepsilon_{TOL}$. Subsequently test LH and RH regions separately with $\varepsilon_{TOLprevious}./2$

If satisfied Yes then accept result and do one Richardson Extrapolation.

If not satisfied No then subdivide into LH and RH regions, then

allocate $\varepsilon_{TOLprevious}/2$ to each region and proceed with Steps#1 and #2 for each

4.7.2 Adaptive Quadrature Algorithm Implementation Issues

Adaptive Quadrature Algorithm Implementation Issues

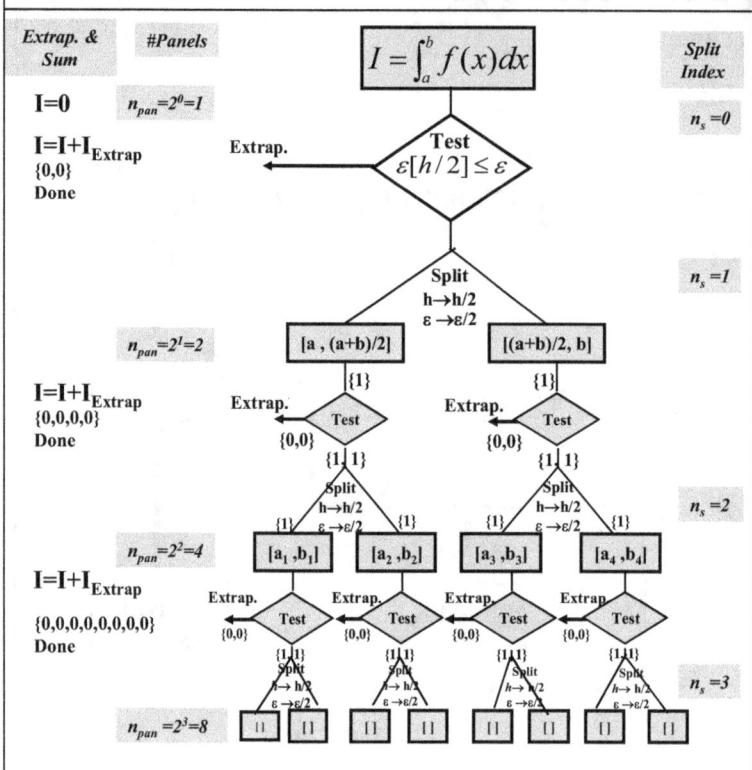

The general structure of the adaptive quadrature algorithm is outlined on the last slide, but there are implementation issues that need to be addressed in order to realize a working algorithm. The main issue is to track the many paths that may occur as the process steps through the i) L/R split, ii) ε_{TOL} test, iii) extrapolate, and iv) split again.

One way to handle this is to carry a set of flags that identifies the active {1} and inactive {0} panels with a concatenated {Left, Right} state {1,1} for the first split and then doubles size with each new split. After the 1st split, the **test** either *extrapolates and outputs* a partial result and continues on to the next split. More specifically, consider the 1st split (n=1) for the case in which the L-panel test passes; we then extrapolate the L-panel , **make it inactive**, and **double**: {0} =>{0 0}; if on the other hand, the R-panel test fails, we split, **leave active**, and **double**: {1} =>{1 1}. The next split state for this case has **four panels** {0 0 1 1} with only 2 active. In this way there are 4 possible resulting states resulting from the n=1 split, viz., i) both L/R pass & extrapolate to yield state {0 0 0 0} and **done**, ii) L pass / R fail yields state {0 0 1 1} iii) L fail / R pass yields state {1 1 0 0}, or iv) both L/R fail yields state {1 1 1 1}. In each case the state for the **2nd split** has 4 panels because we double each active panel {1} and inactive panel {0} after the test; if the *panel state* is {0 0 0 0} processing stops and the integral value is output.

This process continues on the 2nd split where the four states are tested individually; the **inactive panels** states are just doubled, while the **active panels** are each tested to either extrapolate or split again and their states are doubled again. No matter the results are, this new state has 8 panels; the process continues until either the state is all zeros {0 0 0 ... 0} or the desired maximum number of splits, say, $n_{s)max}$ =10 is reached, corresponding to 2^{10} =1024 panels.

The second issue is to trace the coordinate intervals for all panels of each panel in the current split; a sample calculation is given in the slide for n_S=1 & k=1. (Also note that Δ_n halves in each split level n_s .)

$$[a_k, b_k] = [a + (k-1)\Delta_{n_s}, \ a + k\Delta_{n_s}] \quad k = 1,2,\cdots,n_p \qquad \Delta_{n_s} = \frac{a+b}{n_p}; \ n_p = 2^{n_s}$$

4.7.3 Adaptive Quadrature Examples

Adaptive Quadrature Examples

'2*x.*cos(2*x)-(x-2).^2' a=0; b=5;

Itmp	Itmpj	e_tol	# panel	h_{new}	#panels$_{new}$	#panels$_{active}$
0	0	1e-005	1	2.5	2	2
0	0	5e-006	2	1.25	4	4
0	0	2.5e-006	4	0.625	8	8
0	0	1.25e-006	8	0.3125	16	16
0.79355	0.79355	6.25e-007	16	0.15625	32	30
-9.9031	-10.697	3.125e-007	32	0.078125	64	4
-15.306	-5.4032	1.5625e-007	64	0.039063	128	0

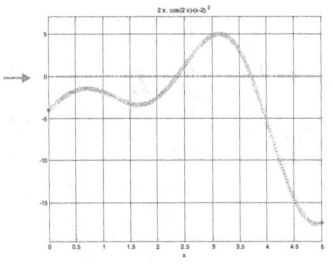

Itmp = -15.3063093510881 ;
e_tol = 1e-005   64 panels

Itmp	Itmpj	e_tol	# panel	h_{new}	#panels$_{new}$	#panels$_{active}$
0	0	1e-007	1	2.5	2	2
0	0	5e-008	2	1.25	4	4
0	0	2.5e-008	4	0.625	8	8
0	0	1.25e-008	8	0.3125	16	16
0	0	6.25e-009	16	0.15625	32	32
0	0	3.125e-009	32	0.078125	64	64
-1.7587	-1.7587	1.5625e-009	64	0.039063	128	10
-15.306	-13.548	7.8125e-010	128	0.019531	256	0

Itmp = -15.3063079957494;
e_tol = 1e-007   128 panels

This is a typical adaptive quadrature example for the function f(x) =2*x.* cos(2*x)-(x-2).^2 over the interval x ε [0,5] (Ex.=7) for two error tolerances e_tol = 10^{-5} and 10^{-7}. The algorithm follows the flow description detailed on the previous slide and is given along with a number of additional examples in the appendix on Slide# 5-21 to 5-22   The plots shows the panels [a$_k$, b$_k$] as red circles overlaid on the function plot (as well tick marks along the x-axis) for the two cases.

The tables show the progression of active panels starting with the original [a,b] panel and then in each split n$_s$ =1,2,3 … until there are no active panels. Also shown for each split n$_s$ are the extrapolation contributions to the integral for Itmpj (col#2) and the cumulative contribution Itmp (col#1), the L/R e_tol (col#3), the #panels (col#4), the new stepsize h$_{new}$ (col#5), and the new panel (col#6), and the active panels (col#7). The final number of active panels in each case is {0 0...0} designated by a simple "0". As we decrease the target tolerance from e_tol = 10^{-5} to 10^{-7} we see (in this case) the number of panels needed increases from 64 to 128. Also note that the values in the cumulative output (col#1) are equal to the value in the (previous split)+ (current split extrapolation) contribution, viz., -15.306 = -1.7587+(-13.548) for the last two rows of the bottom table. Also note that, naturally enough, the contributions come in later splits as e_tol is made smaller. In general, the improvements using the adaptive quadrature are most significant for functions that have large abrupt changes such as the sin(50x)/(π x) of Ex. 9 in the appendix Slide# 5-16 and #5-21.

4.8 *Romberg Integration*

Romberg Integration

- Trapezoidal $(k+1)$-pt down first column $\quad R_{k,1} = I_k^{Trap}[h_k] \ ; \ h_k = \dfrac{b-a}{2^{k-1}}$

- Halving $h_k = h/2^{k-1}$ does not affect constant RO

- Extrapolate across rows to reduce trunc. Error $\quad \varepsilon_{k,1} = -\dfrac{b-a}{12}h_k^2 f''(\xi)$

- Recursion $\quad R_{k,1} = \dfrac{1}{2}\left[R_{k-1,1} + h_{k-1}\underbrace{\sum_{m=1}^{2^{k-2}} f(a+(m-\tfrac{1}{2})h_{k-1})}_{2^{k-2}-\text{fcnal evals.}} \right]$

Trunc. Error Decreases ⟶

	Composite Trapezoidal			
	$O(h^2)$	$O(h^4)$	$O(h^6)$	$O(h^8)$
1-panel	$R_{1,1}[h]$			
2-panels	$R_{2,1}[h/2]$	$R_{2,2}[h]$		
4-panels	$R_{3,1}[h/4]$	$R_{3,2}[h/2]$	$R_{3,3}[h]$	
	⋮			
2^{k-1}-panels	$R_{k,1}[h/2^k]$	$R_{k,2}[h/2^{k-1}]$		$R_{k,k}[h]$

Constant RO Error Indep of h

$h_k = (b-a)/2^{k-1}$

The Romberg integration algorithm computes a sequence of 1-panel, 2-panel, 4-panel, ... composite trapezoidal quadratures to populate the 1st column of the extrapolation table. It can be shown the Romberg integration algorithm has an $\{h^2, h^4, h^6,...\}$ error structure and this allows us to perform Richardson Extrapolation which reduces the truncation error by order h^2 as we proceed across the row representing higher order fits. Moreover, each new row corresponds to a halved stepsize which further reduces truncation error without any penalty because of the constant RO error. This process continues until the truncation error becomes smaller than the RO error; thus Romberg integration allows the integral estimate to approach machine precision in a very effective manner.

The boxed formula gives an efficient recursive method to compute the elements of the 1st column by re-using the previous element in the column. For the purposes of understanding the process it is best to compute the 1-panel, 2-panel, 4-panel, ... composite trapezoidal quadratures directly rather than using the formula. Then each row may be computed using the extrapolation

row#2 $R_{2,2} = R_{2,1}+(R_{2,1}-R_{1,1})/3 = R_{2,1}+\Delta_{row}R_{2,1}/3$

row#3 $R_{3,2} = R_{3,1}+(R_{3,1}-R_{2,1})/3 = R_{3,1}+\Delta_{row}R_{2,1}/3$

   $R_{3,3} = R_{3,2}+(R_{3,2}-R_{22})/15 = R_{3,2}+\Delta_{row}R_{3,2}/15$

4.8.1 Romberg Integration-Examples -1

Romberg Integration-Examples -1

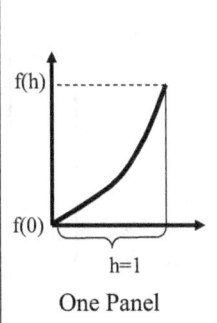

f(h)

f(0)

h=1

One Panel

Ex.#1: $f(x) = x^2$

$$I = \int_0^1 x^2 dx = \frac{x^3}{3}\Big|_0^1 = \frac{1}{3} = .33333\cdots$$

One Panel

$$R_{1,1}[h] = \frac{1}{2}\left[0^2 + 1^2\right] = 1/2 = .50000$$

Two Panels

$$R_{2,1}[h/2] = \frac{1/2}{2}\left[0^2 + 2(1/2)^2 + 1^2\right] = 3/8 = .37500$$

$$R_{2,2}[h] = \frac{4 \cdot R_{2,1}[h/2] - R_{1,1}[h]}{3} = \frac{4(3/8)-1/2}{3} = \frac{1}{3} = .33333$$

$O(h^2)$	$O(h^4)$
.50000 $R_{1,1}[h]$	
.37500 $R_{2,1}[h/2]$	.33333 $R_{2,2}[h]$

Machine Precision!

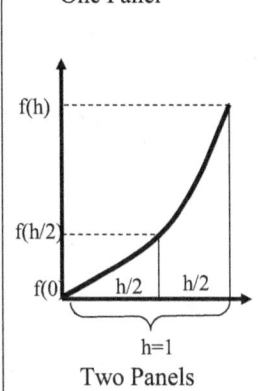

f(h)

f(h/2)

f(0)   h/2   h/2

h=1

Two Panels

Ex. #2: $f(x) = x^3$

$$I = \int_0^1 x^3 dx = \frac{x^4}{4}\Big|_0^1 = \frac{1}{4} = .25000\cdots$$

One Panel

$$R_{1,1}[h] = \frac{1}{2}\left[0^3 + 1^3\right] = 1/2 = .50000$$

Two Panels

$$R_{2,1}[h/2] = \frac{1/2}{2}\left[0^3 + 2(1/2)^3 + 1^3\right] = 5/16 = .31250$$

$$R_{2,2}[h] = \frac{4 \cdot R_{2,1}[h/2] - R_{1,1}[h]}{3} = \frac{4(5/16)-1/2}{3} = 1/4 = .25000$$

$O(h^2)$	$O(h^4)$
.50000 $R_{1,1}[h]$	
.31250 $R_{2,1}[h/2]$	.25000 $R_{2,2}[h]$

Machine Precision!

This simple example, integrating x^2 over [0,1], allows us to verify directly that Richardson extrapolation decreases the truncation error and achieves machine precision. The 1-panel and 2-panel trapezoidal quadratures of the first column $R_{1,1}$ and $R_{2,1}$ represent integration of linear fits which have truncation error $O(h^2)$. Since one extrapolation using these two terms yields $R_{2,2}$ with a truncation error of $O(h^4)$, we expect the quadrature to have zero truncation error and hence the result should reach machine precision as it does. Indeed, we should also obtain zero truncation error integrating x^3 over [0,1] and a simple recalculation yields the exact value of 1/4 to machine precision.

Note that we have not used the recursion formula for this simple problem and also we have not used the numerically efficient form $R_{2,2} = R_{2,1} + (R_{2,1} - R_{1,1})/3$ but used instead the algebraically equivalent form $R_{2,2} = (4R_{2,1} - R_{1,1})/3$ which suffices for our purposes.

4.8.2 Romberg Integration-Examples -2

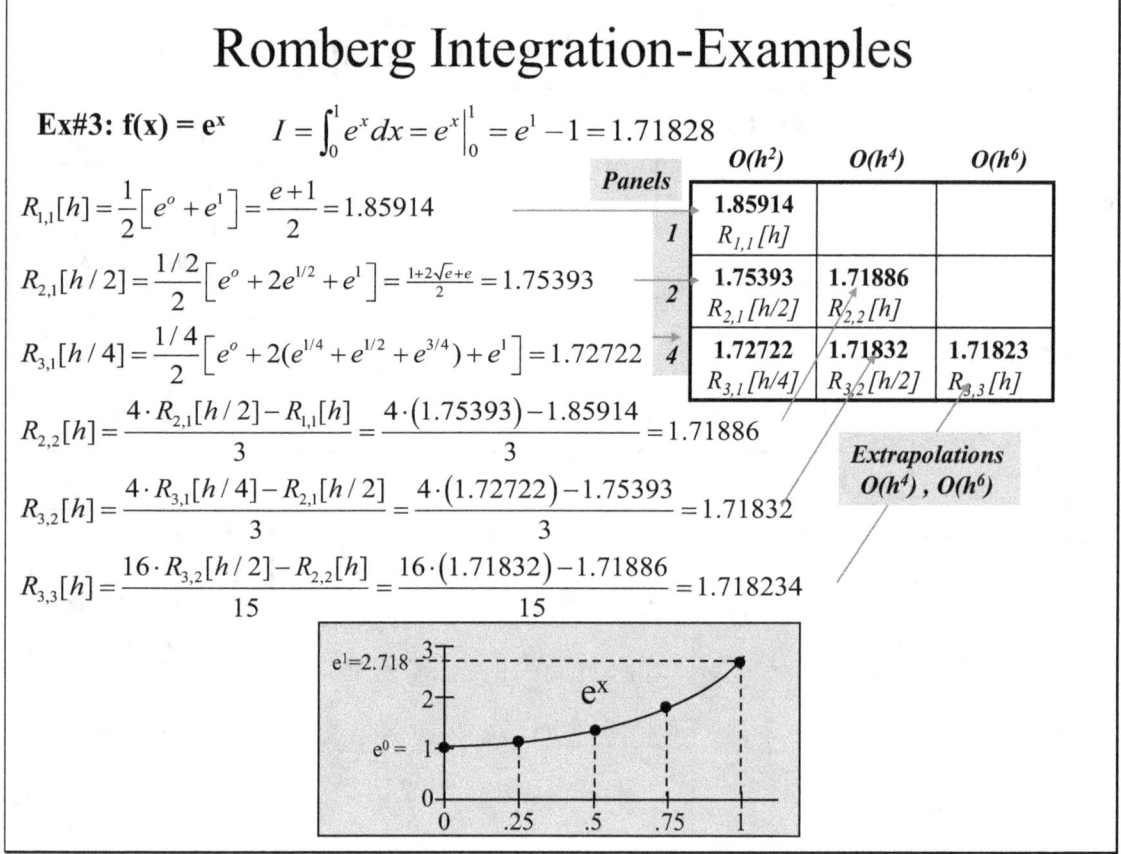

Romberg Integration-Examples

Ex#3: f(x) = e^x $I = \int_0^1 e^x dx = e^x \Big|_0^1 = e^1 - 1 = 1.71828$

$R_{1,1}[h] = \frac{1}{2}\left[e^o + e^1\right] = \frac{e+1}{2} = 1.85914$

$R_{2,1}[h/2] = \frac{1/2}{2}\left[e^o + 2e^{1/2} + e^1\right] = \frac{1+2\sqrt{e}+e}{2} = 1.75393$

$R_{3,1}[h/4] = \frac{1/4}{2}\left[e^o + 2(e^{1/4} + e^{1/2} + e^{3/4}) + e^1\right] = 1.72722$

Panels	$O(h^2)$	$O(h^4)$	$O(h^6)$
1	1.85914 $R_{1,1}[h]$		
2	1.75393 $R_{2,1}[h/2]$	1.71886 $R_{2,2}[h]$	
4	1.72722 $R_{3,1}[h/4]$	1.71832 $R_{3,2}[h/2]$	1.71823 $R_{3,3}[h]$

$R_{2,2}[h] = \frac{4 \cdot R_{2,1}[h/2] - R_{1,1}[h]}{3} = \frac{4 \cdot (1.75393) - 1.85914}{3} = 1.71886$

$R_{3,2}[h] = \frac{4 \cdot R_{3,1}[h/4] - R_{2,1}[h/2]}{3} = \frac{4 \cdot (1.72722) - 1.75393}{3} = 1.71832$

Extrapolations
$O(h^4)$, $O(h^6)$

$R_{3,3}[h] = \frac{16 \cdot R_{3,2}[h/2] - R_{2,2}[h]}{15} = \frac{16 \cdot (1.71832) - 1.71886}{15} = 1.718234$

This second example shows the normal situation in which we need many extrapolations to beat down the truncation error. Since the integrand e^x is not expressible as a finite degree polynomial the truncation error is never exactly zero.

We show the Richardson extrapolation table for three columns leading to truncation error $O(h^6)$ and yielding an entry $R_{3,3}[h]$ that is correct to 4 sd (5 with chopping). Note that the second column has a divisor of 3, while the third column has a divisor of 15; it can be shown that the kth column has a divisor of 4^{k-1} -1. Again, for illustration purposes, we have used a compact expression for the extrapolations instead of the numerically efficient forms that should be used.

4.9 *Gaussian Quadrature*

Gaussian Quadrature

- Quadrature $\qquad I = \int_a^b f(x)dx = \underbrace{\sum_{k=0}^n c_k f(x_k)}_{\text{Wgtd Sum of Nodal Values}}$

  n+1 fixed nodes: $\{x_0, x_1, \cdots x_n\}$

  n+1 arbitrary constants $\quad \{c_0, c_1, \cdots c_n\}$

- Vary the nodes → 2(n+1) arbitrary conditions

  Exact Quadrature for Poly deg $\leq$ 2n+1

	Num Pts per Panel	Num Free Cond.	Poly deg
Trapezoid	2	2	1
Simpson	3	3	3*
Gauss Quad	3	6	5

* Special Symmetry "2" →"3"

Gaussian quadrature is again a weighted sum of the functional values at the n+1 nodes, but instead of choosing a fixed nodal grid with uniform spacing h, the n+1 nodal positions are arbitrary. Moreover, the original integral of the function f(x) over [a,b] is transformed into the integral of a new function g(t) over [-1,1]. Thus in addition to the n+1 arbitrary nodal weighting constants $\{c_0, c_1, \dots c_n\}$ there are arbitrary n+1 nodes $\{t_0, t_1, \dots t_n\}$ in [-1,1] so we have a total of 2(n+1) conditions we can impose on the quadrature solution which makes it exact for a polynomial of degree less than or equal to 2n+1. The table summarizes the characteristics of the Trapezoidal, Simpson and Gaussian quadratures in terms of number of points, number of free conditions, and maximum degree polynomial for exact quadrature (i.e., zero truncation error.) The Trapezoidal method has 2 free conditions and yields an exact quadrature for a polynomial up to degree 1; the Simpson method has 3 free conditions and should therefore only determine a polynomial of degree 2 exactly; however, we get a "bonus" with Simpson because the 1st non-zero contribution to the truncation error is in the x^4 term (not the x^3 term.) so the table entry is exact for degree 3 polynomial. Finally, for the same number of nodes as Simpson, namely 3, the Gauss method has 6 free conditions and leads to exact quadrature for polynomials up to degree 5.

4.9.1 2-pt Gaussian Quadrature

2-pt Gaussian Quadrature

- ## Transformation [a, b] → [-1,+1]

	Gauss-Legendre	
n	Roots(t_k)	Coeff.(c_k)
2	.577350	1.00000
	-.577350	1.00000
3	.774597	.555556
	.000000	.888889
	-.774597	.555556

$$x = \frac{a+b}{2} + \frac{b-a}{2} \cdot t \qquad I = \int_a^b f(x)dx = \int_{-1}^{+1} g(t)dt$$

$$dx = \frac{b-a}{2} \cdot dt \quad \Rightarrow \quad g(t) = \frac{b-a}{2} \cdot f\left(\frac{a+b}{2} + \frac{b-a}{2} \cdot t\right)$$

- ## Evaluation

$$I_G = \int_{-1}^{+1} g(t)dt = \sum_{k=0}^{1} c_k g(t_k) = c_0 g(t_0) + c_1 g(t_1)$$

$$= 1 \cdot g(-1/\sqrt{3}) + 1 \cdot g(+1/\sqrt{3})$$

- ## Error

$$R_T = \begin{cases} 0 & \deg \le 3 \\ \dfrac{g^{(4)}(\xi)}{135} & \deg > 3 \end{cases}$$

The constant 1/135 is derived by integrating $g(t)=t^4$ to find I, forming error $|I\text{-}I_G|$ and noting $g^{(iv)}(t)=24$ as shown at right

$$I = \int_{t=-1}^{+1} t^4 dt = t^5/5 \Big|_{t=-1}^{+1} = 2/5$$

$$I_G = 1 \cdot \left(-1/\sqrt{3}\right)^4 + 1 \cdot \left(1/\sqrt{3}\right)^4 = 2/9$$

$$error = |I - I_G| = |2/5 - 2/9| = 8/45$$

$$= Kg^{(iv)}(\xi) = 24K \Rightarrow K = 1/135$$

The 2-point Gaussian quadrature solution procedure is shown in this slide.
The 1st step is to transform the integral of f(x) over [a,b] into g(t) over [-1,1] ; this is easily done *via* X and new function definition

$$x = (a+b)/2 + (b-a)*t/2 \quad g(t)=(b-a)/2*f(\ (a+b)/2 + (b-a)*t/2\ )$$

The 2nd step is to evaluate the Gaussian quadrature for two points as the weighted sum

$$I_G = c_0\, g(t_0) + c_1\, g(t_1)$$

where the four parameters c_0, c_1, t_0, t_1 must be specified. These nodal points $\{t_0, t_1\}$ and nodal weighting factors $\{c_0, c_1\}$ are read off the n=2 section of the Gauss-Legendre Table. The truncation error, given by the last equation, shows zero error for deg up to 3 and otherwise is equal to $g^{(4)}(\xi) / 135$ for deg > 3.
For a 3-point Gaussian quadrature we would write

$$I_G = c_0\, g(t_0) + c_1\, g(t_1) + c_2\, g(t_2)$$

where the six parameters $c_0, c_1, c_2, t_0, t_1, t_2$ are obtained from the 3-point section of the same a Gauss-Legendre Table. The description so far describes a "somewhat magical" technique and in the next slide we shall demystify the Gaussian quadrature. However the constant 1/135 in the error formula is easily derived by integrating the 4th degree polynomial $g(t)=t^4$ exactly to find I=2/5, computing $I_G = 2/9$ using the 2-pt Gaussian Quadrature table values and forming error $|I\text{-}I_G| = |2/5\text{-}2/9| = K\, g^{(iv)}(\xi)$. Finally, differentiating t^4 to find $g^{(iv)}(t) = 24$ for any value of "t", so $g^{(iv)}(\xi) = 24$, we determine K= (8/45) /24 = 1/135 as shown explicitly on the bottom of the slide.

4.9.2 Gaussian Quadrature – Intuitive Proof

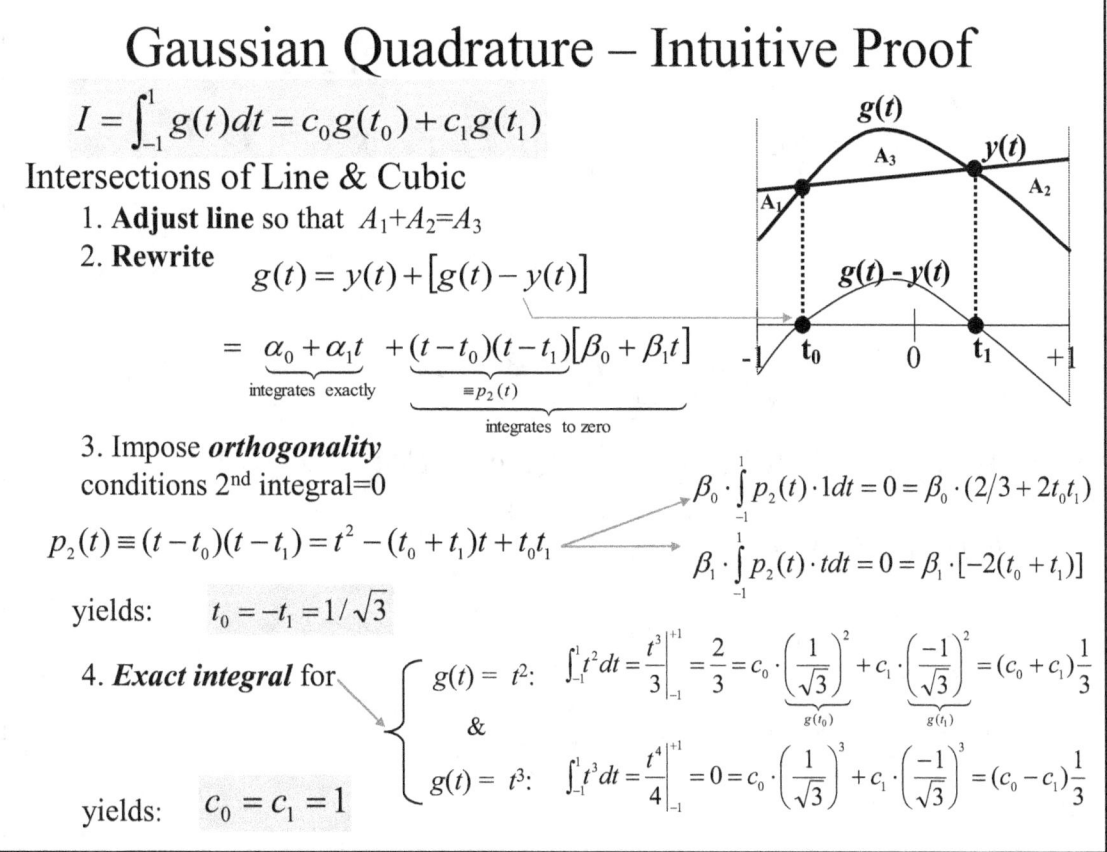

Gaussian Quadrature – Intuitive Proof

$$I = \int_{-1}^{1} g(t)dt = c_0 g(t_0) + c_1 g(t_1)$$

Intersections of Line & Cubic

1. **Adjust line** so that $A_1 + A_2 = A_3$
2. **Rewrite**

$$g(t) = y(t) + [g(t) - y(t)]$$

$$= \underbrace{\alpha_0 + \alpha_1 t}_{\text{integrates exactly}} + \underbrace{(t - t_0)(t - t_1)[\beta_0 + \beta_1 t]}_{\equiv p_2(t)}$$

$$\underbrace{\qquad\qquad\qquad}_{\text{integrates to zero}}$$

3. Impose **orthogonality**
conditions 2nd integral=0

$$p_2(t) \equiv (t - t_0)(t - t_1) = t^2 - (t_0 + t_1)t + t_0 t_1$$

$$\beta_0 \cdot \int_{-1}^{1} p_2(t) \cdot 1 dt = 0 = \beta_0 \cdot (2/3 + 2t_0 t_1)$$

$$\beta_1 \cdot \int_{-1}^{1} p_2(t) \cdot t dt = 0 = \beta_1 \cdot [-2(t_0 + t_1)]$$

yields: $\quad t_0 = -t_1 = 1/\sqrt{3}$

4. **Exact integral** for

$$g(t) = t^2: \quad \int_{-1}^{1} t^2 dt = \frac{t^3}{3}\Big|_{-1}^{+1} = \frac{2}{3} = c_0 \cdot \underbrace{\left(\frac{1}{\sqrt{3}}\right)^2}_{g(t_0)} + c_1 \cdot \underbrace{\left(\frac{-1}{\sqrt{3}}\right)^2}_{g(t_1)} = (c_0 + c_1)\frac{1}{3}$$

&

$$g(t) = t^3: \quad \int_{-1}^{1} t^3 dt = \frac{t^4}{4}\Big|_{-1}^{+1} = 0 = c_0 \cdot \left(\frac{1}{\sqrt{3}}\right)^3 + c_1 \cdot \left(\frac{-1}{\sqrt{3}}\right)^3 = (c_0 - c_1)\frac{1}{3}$$

yields: $\quad c_0 = c_1 = 1$

In order to understand how and why Gaussian quadrature works we consider a 2-point (n=1) quadrature and analyze the integral of the cubic curve shown over the interval [-1,1]. Consider the straight line $y(t) = \alpha_0 + \alpha_1 t$ intersecting the cubic curve $g(t)$ at the two points shown in the figure. Now "move/rotate" $y(t)$ so that the area under it equals the area under $g(t)$. This will be true provided that the areas shown satisfy $A_1 + A_2 = A_3$.

Formally rewriting $g(t)$ as $y(t)$ plus the cubic $[g(t)-y(t)]$, it is clear from the figure that the cubic function in the square brackets is zero at both nodes t_0 and t_1 and this allows us to re-write it as the product of $(t-t_0)(t-t_1)$ and a linear function $(\beta_0+\beta_1 t)$. Moreover, the straight line $y(t)$ was chosen so that the area under it is exactly the desired integral of $g(t)$; hence the integral of the cubic function $[g(t)-y(t)] = (t-t_0)(t-t_1)\cdot(\beta_0+\beta_1 t)$ must be is zero for arbitrary values of β_0 and β_1.

Imposing zero contributions for $\beta_0 \cdot(t-t_0)(t-t_1)$ and $\beta_1 \cdot t(t-t_0)(t-t_1)$ integrated over [-1,1] yields two equations (i) $t_0 + t_1 = 0$ and (ii) $t_0 \cdot t_1 = -1/3$; thus the two unknown nodes are $t_0 = -t_1 = 3^{-1/2}$.

Requiring exactness for the 2-point quadrature $I_G = c_0 g(t_0) + c_1 g(t_1)$ when integrating the monomials $g(t)=t^2$ and $g(t)=t^3$. Thus performing these exact integrals gives 2/3 and 0 respectively and upon comparing these values to the quadrature formula results, we find $c_0 = c_1 = 1$. Thus, we have reproduced the nodes and coefficients of the 2-point Gaussian quadrature.

Since the difference function $g(t) - y(t)$ integrates to zero, we may apply the Gauss quadrature to $y(t)$ itself to yield $I_G = c_0 y(t_0) + c_1 y(t_1) = 1\cdot(\alpha_0 + \alpha_1 t_0) + 1\cdot(\alpha_0 + \alpha_1 t_1) = \alpha_0 (1 + 1) + \alpha_1(t_0 + t_1) = 2\alpha_0 + 0$.

Also note that direct integration of the linear function $y(t) = \alpha_0 + \alpha_1 t$ over [-1,1] yields $2\alpha_0$ as it must.

4.9.3 n-pt Gaussian Quadrature

n-pt Gaussian Quadrature

- **Integral & Error** $I = \int_{-1}^{+1} g(t)\,dt = I_G + R_n$

$$I_G = \sum_{k=0}^{n-1} c_k g(t_k) \qquad\qquad c_k = \frac{2}{(1-t_k^2)\cdot[P_n'(t_k)]^2}$$

$$R_n = \frac{g^{(2n)}(\xi)}{(2n)!}\cdot\left(\frac{2}{2n+1} - \sum_{k=0}^{n} c_k t_k^{2n}\right) \quad \deg > 2n-1$$

- t_k are zeros of nth degree Legendre Poly $P_n(t_k) = 0$

$$P_0(t) = 1 \;;\; P_1(t) = t$$

$$P_2(t) = \frac{1}{2}(3t^2 - 1) \;;\; t_0 = -\frac{1}{\sqrt{3}} \;;\; t_1 = +\frac{1}{\sqrt{3}}$$

$$P_3(t) = \frac{1}{2}(5t^3 - 3t) \;;\; t_0 = 0 \;;\; t_1 = -\sqrt{\frac{3}{5}} \;;\; t_2 = +\sqrt{\frac{3}{5}}$$

The Integral I_G, coefficients c_k, nodes t_k, and truncation error R_n for an n-point Gaussian quadrature is given on this slide. The "arbitrary" node locations t_k turn out to be the zeros of the Legendre polynomial functions of degree n, $P_n(t)$. The first four Legendre Polynomials are given on the slide and the zeros of these polynomials are easily seen to be $\{-1/(3)^{1/2}, \ 1/(3)^{1/2}\}$ for $P_2(t)$ and $\{0, \ -(2/5)^{1/2}, \ (2/5)^{1/2}\}$ for $P_3(t)$.

Note that the recursion relations $(k+1)\cdot P_{k+1} = (2k+1)\cdot x\cdot P_k - k\cdot P_{k-1}$ can be used to generate the polynomials "on the fly" (see Slide# 3-8).

4.9.4 Other Orthonormal Quadratures

Other Orthonormal Quadratures

$$I = \int_a^b f(x)dx \Rightarrow \int_{t_a}^{t_b} w(t)g(t)dt = \sum c_k g(t_k) = I_G$$

Name	t_a	t_b	$w(t)$	Zeros of	c_k
Hermite	$-\infty$	$+\infty$	e^{-t^2}	$H_m(t)$	Table
Laguerre	0	$+\infty$	e^{-t}	$L_m(t)$	Table
Tschebyshev	-1	+1	$(1-t^2)^{-1/2}$	$T_m(t)$	π/m
Legendre	-1	+1	1	$P_m(t)$	Table

Recursion Relations

First Few Polynomials

Hermite $\quad H_{k+1} = 2xH_k - 2kH_{k-1}$
$\qquad H_0 = 1; H_1 = 2x ; H_2 = (4x^2 - 2) ; H_3 = (8x^3 - 12x)$

Laguerre $\quad L_{k+1} = (-x + 2k + 1)L_k - kL_{k-1}$
$\qquad L_0 = 1; L_1 = x; L_2 = (x^2 - 4x + 2)/2$
$\qquad L_3 = (-x^3 + 9x^2 - 18x + 6)/6$

Tschebyshev $\quad T_k = 2xT_k - T_{k-1}$
$\qquad T_0 = 1; T_1 = x ; T_2 = (2x^2 - 1) ; T_3 = (4x^3 - 3x)$

Lagrange $(k+1)P_{k+1} = (2k+1)xP_k - kP_{k-1}$
$\qquad P_0 = 1; P_1 = x ; P_2 = (3x^2 - 1)/2; P_3 = x(5x^2 - 3)/2$

The Gaussian quadrature method has been developed for the Legendre polynomials but it is a general technique and may be applied to other orthonormal polynomials such as Hermite, Laguerre, and Tschebyshev. The main difference is that new transformations are needed to convert the integral of $f(x)$ over [a,b] to a $g(t)$ over $[t_a, t_b]$ as indicated in the table for each polynomial type. Also the appropriate weighting function $w(t)$ must appear in the integrand of the integration over the variable "t". It is crucial to factor out the weighting function and whatever remains is the new $g(t)$ function that is used in the Gaussian quadrature, *e.g.*, for 3-points we have

$I_G = c_0\, g(t_0) + c_1\, g(t_1) + c_2\, g(t_2)$

even though the integral contains the product $w(t)*g(t)$. The weighting factor $w(t)$ can be considered as part of the integral structure rather than the integrand.

4.9.5 Orthonormal Quadratures - Example

Orthonormal Quadratures - Example

$$I = \int_0^1 \frac{x}{\sqrt{4x(1-x)}}\, dx$$

Quadrature Equations

Map: $\quad x = \dfrac{t+1}{2} \;\; ; \;\; dx = \dfrac{1}{2} dt$

$$I = \int_{-1}^{1} \frac{(t+1)/2}{\sqrt{4\frac{(t+1)}{2}\left(1-\frac{(t+1)}{2}\right)}} \frac{dt}{2} = \frac{1}{4}\int_{-1}^{1} \frac{(t+1)}{\sqrt{1-t^2}} dt = \frac{1}{4}\int_{-1}^{1} \sqrt{\frac{1+t}{1-t}}\, dt$$

		Gauss-Legendre	
	n	Roots(t_k)	Coeff.(c_k)
	2	.577350	1.00000
		-.577350	1.00000
	3	.774597	.555556
		.000000	.888889
		-.774597	.555556

Gauss-Legendre:

$$I_{G-L} = \frac{1}{4}\left(.55556\left\{\sqrt{\frac{1+.774597}{1-.774597}} + \sqrt{\frac{1-.774597}{1+.774597}}\right\} + .88889\sqrt{\frac{1+0.0}{1-0.0}}\right) = .661426$$

		Gauss-Tschebyshev	
	n	Roots(t_k)	Coeff.(c_k)
	2	.707106	$\pi/2$
		-.707106	$\pi/2$
	3	.866025	$\pi/3$
		.000000	$\pi/3$
		-.866025	$\pi/3$

Gauss-Tschebyshev:

$$I_{G-Tch} = \frac{1}{4}\cdot\frac{\pi}{3}\left((1+.866025) + (1+0.0) + (1-.866025)\right) = \frac{\pi}{4} = .785398$$

Exact!

$$T_3 = (4x^3 - 3x) = 0 \quad \Rightarrow x = 0 \; ; \; x = \pm\sqrt{3/4} = \pm.866025$$

The integral given is over [0,1] and so must first be put into the proper form of an integral over [-1,1] to perform Gauss-Legendre quadrature. Upon making the transformation we find two usable forms of the integrand $(t+1)/(1-t^2)^{1/2}$

$$g(t) = [(1+t)/(1-t)]^{1/2} \quad w(t) = 1 \qquad \text{for Gauss-Legendre}$$
$$g(t) = (1+t) \qquad w(t) = 1/(1-t)^{1/2} \quad \text{for Gauss-Tschebyshev}$$

For (i) we apply the 3-point Gauss-Legendre Table coefficients and nodes to the "whole" function in the integrand (we just factored out a common term in the numerator and denominator). The weighting factor $w(t) = 1$.

For (ii) we apply the 3-point Gauss-Tschebyshev Table coefficients to the function $g(t) = (1+t)$ and only $(t+1)$ appears in I_{G-Tch} evaluations. The weighting function $w(t) = 1/(1-t)^{1/2}$ is part of the integral and not part of the integrand!

This Gauss-Tschebyshev result is more accurate because the weighting function factors out the singular part of the integrand and makes it part of the integration structure where it does no harm.

Also note the Tschebyshev coefficients can be factored out of the weighted quadrature sum since they are the same for each term in the sum. They are $\pi/2$ for 2-point, $\pi/3$ for 3-point and π/n for n-point. This is related to their uniform structure over the interval [-1,1] which was previously used to advantage in Tschebyshev economization.

4.9.6 Multi-panel Gauss Quadratures

Multi-panel Gauss Quadratures

n_p Gauss Panels
from n_s L/R splits

$$n_p = 2^{n_s} \; ; \; h_p = (b-a)/2^{n_s} = (b-a)/n_p$$

$$a_k = a + (k-1) \cdot h_p \; ; \; b_k = a + k \cdot h_p$$

Example n_s=2, n_p=4

1 panel: h = b - a = 6
4 panels: h_p =h/4 =(7-1)/4 =3/2

The red positions in each h/4 panel
are given by

$$x^k(u_i) = a + (h_p/2)[u_i + (2k-1)]$$

Gauss roots, $u_i \in (-1,1)$ i=1,..., n_g
Panels are k=1,2,3,4

for n_g=2, i=1,2, k=3: roots u_i=±.57735
$x^3(u_2)$=1+(1.5/2)(+.57735+2(3)-1)
= 5.1380, 4.3170

Transform each
Gauss Panel from
[a_k ,b_k] to [-1, 1]
k=1,2,... n_p

$$h_1^k = (b_k - a_k)/2 = h_p/2 \text{ ("indep of } k\text{")}$$

$$h_2^k = a + (2k-1) \cdot h_p/2$$

$$x^k(t) = h_1^k \cdot t + h_2^k = (h_p/2)[t + (2k-1)] + a$$

$g^k(t)$ for
k^{th} panel

$$g^k(t) = h_1^k f(x^k(t)) = a + (h_p/2) \cdot f[(h_p/2)[t + (2k-1)]$$

concatenate
all panels

$$G(K,T) = (h_p/2) \cdot F[(h_p/2) \cdot \{T + (2K-1)\} + A]$$

All are $n_p \times n_g$ Matrices

$$T =: \begin{matrix} 1 \\ \vdots \\ n_p \end{matrix} \begin{bmatrix} t_1 & t_2 & \cdots & t_{n_g} \\ \vdots & & & \vdots \\ t_1 & t_2 & & t_{n_g} \end{bmatrix} \; ; \; (2K-1) = \begin{bmatrix} 2k_1-1 & 2k_2-1 & \cdots & 2k_{n_g}-1 \\ \vdots & & & \vdots \\ 2k_1-1 & 2k_2-1 & & 2k_{n_g}-1 \end{bmatrix} \; ; \; A = \begin{bmatrix} a & a & \cdots & a \\ \vdots & & & \vdots \\ a & a & & a \end{bmatrix} \; ; \; C = \begin{bmatrix} c_1 & c_2 & \cdots & c_{n_g} \\ & & & \\ c_1 & c_2 & & c_{n_g} \end{bmatrix}$$

Gauss roots   *2K-1 Matrix*   *Const. Matrix*   **Gauss Coeff.**

Form Gauss-Legendre
Quadrature using
" .*" matrix product

$$I_G(n_p) = Sum(Sum(C \cdot {}^*G))$$
rows   cols

The multi-panel Gauss quadrature allows for a trade-off between the number of Gauss roots n_g and the number of panels $n_p = 2^{ns}$ where n_s is the number of splits h→h/2. The result here is given in terms of a point by point (.*) product of two ng x np matrices (C.*G). The Gauss-Legendre coordinate transformation maps the original interval [a,b] to [-1,1] *for each panel via*

$$x^k(t) = h_1 t + h_2^k,$$

where h_1=h_p/2 is independent of k. This leads to a known pattern for each panel k=1,2,3,...,n_p (=2^s) and the very compact matrix formulation.

The example in the upper right box illustrates the case for *2 splits* yielding 4 panels (np=2^{ns}=2^2) over the interval [a, b] = [1, 7] and yields 4 panels of equal width h_p =(b-a)/4 =(7-1)/4 = 3/2. We readily write down the individual panels as panel#1: [a,a+h_p]; panel#2: [a+h_p,a+2 h_p]; panel#3 covers [a+2h_p,a+3h_p]; and panel#4: [a+3h_p,a+4h_p].

In the general case of n_p panels we compute the transformations of each panel by first writing down the intervals for each as [a_k, b_k] = [a+(k-1)*h_p , a+k*h_p] which yields a constant scale factor h_1^k= (b_k-a_k)/2 = h_p /2 and a k-dependent displacement h_2^k=(b_k+a_k)/2 = a + (2k-1)*h_p/2

Thus, the transformation takes the form

$$x^k(t) = h_1^k t + h_2^k = (h_p/2)*[t + (2k-1)] + a$$

which scales and displaces the fixed Gauss-Legendre roots located in the interval t = u_k ∈ (-1, 1) so that they are properly located for the function evaluation f(x(t)) in each of the n_p panels. The next step concatenates the n_p rows of panels and n_g columns of Gauss roots to form n_p x n_g function matricies **F** and **G** and use a point-by-point(.*) multiplication of **G** with the Gauss coefficient matrix **C**, viz., **G.*C** . Finally, a sum over columns to yields a vector of panel contributions and a sum over rows of panel contributions yields the multi-panel Gauss estimate of the integral. See Appendix Slides# 5-21 to 5-27 "Gauss 1d Multi-Panel Quadrature" .

4.9.7 Multi-Panel Gauss Quadrature Examples

Multi-Panel Gauss Quadrature vs. Simpson

N_gauss

N_feval(Ng,Np)

10	30	70	150
9	27	63	135
8	24	56	120
7	21	49	105
6	18	42	90
5	15	35	75
4	12	28	60
3	9	21	45
2	6	14	30

N_panels

Adaptive Simpson Processing

itmp	itmpj cor	e_tol	prev.#panels	h	new #pnls	active pnl
0	0	0.0001	1	1	2	2
0	0	5e-005	2	0.5	4	4
0	0	2.5e-005	4	0.25	8	8
-3.2803	-3.2803	1.25e-005	8	0.125	6	12
-19.733	-16.453	6.25e-006	16	0.0625	32	10
-4.4451	15.288	3.125e-006	32	0.03125	64	16
-1.426	3.0191	1.5625e-006	64	0.015625	128	0

Adaptive_Simp	n_panel Simpson	npanel = Nct
-1.42604625428452	-1.42601563466775	92

Multi-Panel fixed n-pt Gaussian Ng =10   (2 methods)

ns	npanel	I_g1	I_g2
0	1	-1.42587440747554	-1.42587440747554
1	2	-1.42602485069542	-1.42602485069543
2	4	-1.42602474882599	-1.42602474882599
3	8	-1.42602475453381	-1.42602475453382

Adaptive Simpson Pts

↓Ng / Np→	1	2	4	8
↓ 10	-1.42587440747554	-1.42602485069543	-1.42602474882599	-1.42602475453382
9	-1.42474119101412	-1.42602537713113	-1.42602475654225	-1.42602475634627
8	-1.42062985259415	-1.42601651364456	-1.42602476373793	-1.42602475634681
7	-1.42709257843547	-1.42651124416141	-1.42634333462073	-1.42600995093497
6	-1.59811707838187	-1.4240321817078	-1.42601824419742	-1.42602475736169
5	-2.37944845625385	-1.43359468513443	-1.42581597326214	-1.42602421089333
4	2.2128583078819	-1.68818951503488	-1.42673748422969	-1.42599224728376
3	10.7423546062407	-2.25529308766008	-1.53750762381997	-1.42785096829333
2	23.3997800502012	8.58507141993743	-1.5043471382549	-1.5087976316098

Nfeval_S	Adapt_Simp	n-panel_Simp	1-pnl_Gauss	multi-pnl_Gauss	Nfeval_G
92	-1.42604625428452	-1.42601563466775	-1.42587440747554	-1.42602475453381	150

The plots and tables for the case $f(x)=(100./x.\^2).*sin(10./x)$ $x\in[\,1,3\,]$ e_tol=10^{-4} are given for both Adaptive Simpson and Multi-panel Gauss in this slide.

Mutlti-Panel Gauss Quadrature: We compute a matrix of Multi-panel Gauss quadratures for n_p #panels and n_g #Gauss points. The Gauss-Legendre roots and coefficient tables provide the matrices **C** and **G** is evaluated at appropriate values in each panel as discussed on the previous slide. For a fixed Gauss quadrature order n_g, the panel halving iterations are stopped when the relative error between to successive splits (n_s), $\Delta I_G \leq$ e_tol =10^{-4} (say). Increasing #Gauss pts to n_g yields a polynomial accuracy of **$(2*n_g-1)$** and incrementing n_s reduces the panel size to $h_p=(b-a)/2^{ns}$. Thus the total error is given by the number of panels, n_p times the Gauss polynomial error for a single panel of width h_p

$$\varepsilon_{tot} = 2^{n_s} \cdot \left\{ \gamma(n_g,n_s) \cdot \frac{h_p}{2} \cdot f^{(2 \cdot n_g)}(x^{(k)}(u_i);\xi) \cdot \frac{h_p^{2 \cdot n_g}}{(2 \cdot n_g)!} \right\}_{max} \leq \gamma(n_g,n_s) \cdot M \cdot \frac{(b-a)^{2n_g+1}}{2^{2n_g n_s+1}}$$

where M is the max derivative in ξ ε [a,b] and $\gamma(n_g,n_s)$ needs to be determined for each Gauss case. The matrix $I_G(n_p,n_g)$ yields a the (9 x 4) table placed just below the interpolated MatLab® contour plot which gives a visual representation of the (Np,Ng) convergence regions; the number of functional evaluations, "Nfevals" is placed left of the contours. The 36 table values are marked as **red circles** surrounding *. [Note that only the [1-,2-,4-,8- panel cases are captured by the halving splits.]. The panel halvings continue until the e_tol = 10^{-3} is met for each n_g-point Gauss case; the most accurate stop value is at the upper right [n_p, n_g]=[8,10] and the error increases as we sweep down to [n_p, n_g]=[1,2] bottom left corner. Because we evaluate Gauss for all the previous previous "splits", the number of panels processed is actually $(2^{nconv+1}-1)$ and thus Nfeval =$(2^{ns+1}-1)*n_g$; for the current example n_s=3, n_g=10 yields $(2^{(3+1)}-1)*10$ =**150**.

Adaptive Simpson Quadrature: The plot at the bottom right shows the **92 function evaluation points** for the Adaptive Simpson method and the Simpson processing table illustrates how it converges when the **#active panels becomes "0"**.

Comparison Multi-Panel Gauss vs Adaptive Simpson: Below the Gauss table is a small table which shows that the "multi-pnl Gauss" requires 150 Nfevals versus 92 for "Adapt Simp" and "n-panel simp".

4.9.8 Integration Methods Comparison

Integration Methods Comparison

Method / Property	Gaussian Quad.	Simpson Composite	Trapezoid Composite
Num. of nodes *(Fixed Polynomial Accuracy)**	m	2m	4m
Computational Load (Relative)	1	2	4
Additional Pts (accuracy)	Recompute t_k, c_k	Easy - Re-use previous fcn evals	Easy - Re-use previous fcn evals
Nodal Spacing	Fcn Zeros: $P_n(t_k)=0$	Evenly spaced nodes	Arbitrarily spaced nodes
Special considerations	Map [a,b] $\rightarrow$[-1,1] Multi-panel Fcn Evals have Pattern	n = even Increase by two	n = even/odd Any number

* See Reference 7 p. 250 for case studies leading to this "rule-of-thumb"

The three basic methods Gaussian quadrature, Simpson Composite and Trapezoid Composite methods are compared. Variations of these methods using adaptive stepsize control and Richardson extrapolation extend their utility. The table specifically compares (i) the number of nodes needed for a fixed degree of polynomial accuracy, (ii) relative computational load (functional $f(x_k)$ and weight c_k evaluations), (iii) the ease with which additional points may be added, (iv) the nodal spacing and (v) any other additional considerations that may affect their utility.

A Gaussian with "m nodes" (1+ "degree index n") yields zero truncation error for polynomials of degree 2m-1, while for the same degree of exactness, the Simpson method requires twice as many or 2m nodes and the trapezoid method requires 4 times as many or 4m nodes. If we assume that dominant computational load is in the functional evaluations (weighting coefficient computations are generally of no consequence), then the computational load increases from 1 for Gaussian to 2 for Simpson to 4 for Trapezoid.

Adding new points is quite easy for both Simpson and Trapezoid and they re-use previous functional evaluations, but Gauss requires re-computation of the zeros of a higher degree polynomial and new coefficients when we add additional points.

Nodal spacing is important when we need to adapt to rapidly varying functions and Simpson requires sets of evenly spaced nodes to form its three point panels while Trapezoidal only uses two points so all nodes can be arbitrarily spaced. The nodes for Gauss are determined by the Legendre zeros and may not be changed at all. Also for Gauss, the integration interval is fixed and requires a mapping transformation from [a,b] to [-1,1] which is an extra burden.

4.10 Improper Integral Techniques

Improper Integral Techniques

1. **End point singularity of integrand**

$$I_1 = \int_a^b \frac{g(x)}{(x-a)^p}dx$$

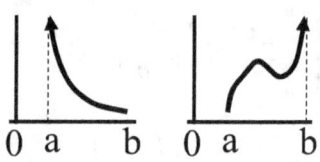

(convergent for $0 < p < 1$)
($p \leq 0$ no singularity!)

2. **Infinite limits of integration**

$$I_2 = \int_{x=a}^{\infty} f(x)dx$$

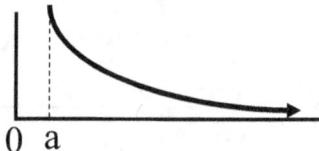

- **Methods to make proper**
   - Absorb singular point into integration variable
   - Invert coord: $x \rightarrow t^1$ changes $x = \inf \rightarrow t = 0$
   - Integrate by parts
   - Polynomial approx. + integrate $g(x)-P_n(x)$ for small correction
   - Weight $w(t)$ is convergence factor for orthonormal polynomial
   - "Excise" end point (open Newton-Cotes)

Improper integrals have integrands that diverge at the end points or they have infinite limits. If the integrand is written as $g(x) / (x-a)^p$, where $g(x)$ is continuous and has no singularities in [a,b] then provided $0<p<1$ the improper integral converges. One can easily show that the integral of $1/(x-a)^p$ is $(x-a)^{1-p}/(1-p) |_a^b = (b-a)^{1-p}/(1-p) - 0$ converges; an integration by parts will show that the integral of $g(x) / (x-a)^p$ converges as well

To evaluate an integral numerically, we must first know that it converges analytically, *i.e.*, that it is a proper integral. There are a number of methods to make an convergent improper integral proper. They include absorbing the singular point into the integration variable, inverting the coordinate, integrating by parts, and showing the main polynomial contribution converges as $N \rightarrow \infty$ and the remaining integral gives a small correction. We can also use an orthonormal function whose weighting factor is the offending part of the integrand, or we can excise the end point and use an open Newton-Cotes method. Several examples are given in the sequel.

4.10.1   Improper Integrals –Examples

Improper Integrals –Examples

- **End point**

$$I = \int_{x=0}^{x=b} \frac{f(x)}{\sqrt{x}} dx \qquad \text{let } x = t^2 \quad dx = 2t\,dt$$

$$I = \int_{t=0}^{t=\sqrt{b}} \frac{f(t^2)}{\sqrt{t^2}} 2t \cdot dt = 2 \int_{t=0}^{t=\sqrt{b}} f(t^2)\,dt$$

- **Trig Subs**

$$\text{let } x = \cos t \quad dx = -\sin t \cdot dt$$

$$x = -1 \rightarrow t = \pi$$

$$x = +1 \rightarrow t = 0$$

$$I = \int_{x=-1}^{x=+1} \frac{f(x)}{\sqrt{1-x^2}} dx$$

$$I = \int_{t=\pi}^{t=0} \frac{f(\cos t)}{\sqrt{1-\cos^2 t}} (-\sin t \cdot dt) = \int_{t=0}^{t=\pi} f(\cos t)\,dt$$

- **Infinite Limit**

$$I = \int_{x=1}^{x=\infty} \frac{e^{-x}}{x^2} dx \qquad \text{let } x = y^{-1} \quad dx = -y^{-2} \cdot dy$$

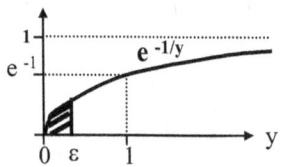

$$I = \int_{y=0}^{y=1} e^{-1/y} dy = \underbrace{\int_{y=0}^{y=\varepsilon=10^{-4}} e^{-1/y}\,dy}_{<10^{-4}\cdot\exp(-1/10^{-4})} + \int_{y=\varepsilon=10^{-4}}^{y=1} e^{-1/y} dy = .1485$$

A change of variables often eliminates a singularity in the integrand as in this first example where the substitution $x = t^2$ gives a factor of t in the differential to cancel the offending term $(x)^{1/2} = (t^2)^{1/2}$ in the denominator.

In the second example, an obvious trigonometric substitution $x = \cos(t)$ gives a factor of $\sin(t)$ in the differential which cancels the offending $(1-x^2)^{1/2} = (1-\cos(t)^2)^{1/2}$ term in the denominator.

In the third example, an infinite limit is mapped to zero by the transformation $x=1/y$ and the resulting integral is broken up into one that approaches zero and one that is integrable for all non-zero lower limits.

4.10.2   Improper Integrals with End Point Singularities

Improper Integrals with End Point Singularities

- Improper Integral of form:

  $p \leq 0$ not improper $(x-a)^{|p|} g(x)$

  $p > 1$ integral diverges

  $$I = \int_{x=a}^{x=b} \frac{g(x)}{(x-a)^p} dx \quad 0 < p < 1$$

- Break up in Main Polynomial Contribution and a Small Correction

  $$I = \int_{x=a}^{x=b} \frac{P_{n>p}(x-a)}{(x-a)^p} dx + \int_{x=a}^{x=b} \frac{\left[g(x)-P_p(x-a)\right]}{(x-a)^p} dx$$

  $\underbrace{\phantom{I = \int_{x=a}^{x=b} \frac{P_{n>p}(x-a)}{(x-a)^p} dx}}_{\substack{\text{I}_1\text{: Main Contribution} \\ \text{Exactly Integrable Polynomial}}}$ $\underbrace{\phantom{\int_{x=a}^{x=b} \frac{\left[g(x)-P_p(x-a)\right]}{(x-a)^p} dx}}_{\substack{\text{I}_2\text{: Small Correction} \\ \text{Composite Simpson}}}$

  Expand Poly abt x=a Degree n > p to cancel denom. $(x-a)^p$

  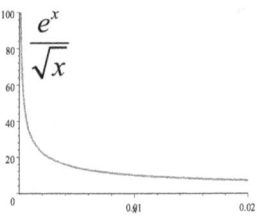

  $\dfrac{e^x}{\sqrt{x}}$

- Example: $a=0$, $p=1/2$, $g(x)=e^x$

  Poly n=4 > p=1/2 about x=a =0

  $$I = \int_{x=0}^{x=1} \frac{e^x}{\sqrt{x}} dx \qquad P_4(x) = 1 + x + \frac{x^2}{2} + \frac{x^3}{3!} + \frac{x^4}{4!} \qquad G(x) = \begin{cases} \dfrac{e^x - P_4(x)}{\sqrt{x}} & x \neq 0 \\ 0 & x = 0 \end{cases}$$

  Note: *We define difference fcn G(x)=0 at the singularity*

  $$G(x) = \frac{\dfrac{x^5}{5!} + \dfrac{x^6}{6!} + \cdots}{\sqrt{x}} = \frac{x^{9/2}}{5!} + \frac{x^{11/2}}{6!} + \cdots \text{ for } x \neq 0$$

This improper integral with integrand $g(x)/(x-a)^p$, converges for all exponents $0 < p < 1$ provided that $g(x)$ is continuous, has no other singularities in $[a, b]$. One approach is to first expand $g(x)$ as a Taylor polynomial of degree n, $P_n(x)$ about $x = a$ so as to cancel the offending $(x-a)^p$ in the denominator; this polynomial can be integrated analytically to provide the main contribution to the original integral. Then, in order to improve this approximate quadrature, the truncation error of the Taylor expansion is divided by $(x-a)^p$ to define a function $G(x)= [g(x)-P_n(x)]/(x-a)^p$ which needs to be integrated over the limits $[a, b]$ using a numerical technique. Although $G(x)$ is not well behaved at $x=a$, the term $G(x)$ is known to be small and we can arbitrarily set $G(a) = 0$ to make $G(x)$ well behaved everywhere.

This method approximates the integral as a "main contribution" from the Taylor polynomial and adds a small correction by numerically integrating the function $G(x)$. Since the integral is known to converge, arbitrarily defining $G(a) = 0$ at the single point $x = a$, has little effect on the computation of an already small correction. Geometrically a small area under the integrand in the region near the asymptote at $x=a$ is effectively dropped by forcing $G(a)=0$.

Taking, for example, an integrand $e^x / x^{1/2}$, corresponding to $g(x)= e^x$, $a=0$ and $p=1/2$, we can write down the Taylor polynomial $P_4(x)$ and integrate it analytically. Further defining $G(x) = [x^5/5! + x^6/6! + x^7/7! + ...] / x^{1/2}$ and setting $G(0)=0$, we are set to compute the small correction integral using any numerical quadrature technique in our "toolbox". The next slide applies a 2-panel composite Simpson quadrature technique to the correction integral.

4.10.3   **Simpson Estimate of Improper Integral**

Simpson Estimate of Improper Integral

Main Contribution: *Analytic Integration of* $P_4(x) / x^{1/2}$

$$I_1 = \int_{x=0}^{x=1} \frac{P_4(x)}{\sqrt{x}}\, dx = \left[\frac{x^{1/2}}{1/2} + \frac{x^{3/2}}{3/2} + \frac{x^{5/2}}{2\cdot 5/2} + \frac{x^{7/2}}{6\cdot 7/2} + \frac{x^{9/2}}{24\cdot 9/2} \right]_0^1 = 2.9235450$$

Small Correction: *2-Panel Simpson of G(x)*

$$I_2 = \underbrace{\int_{x=0}^{x=1} \frac{\left[e^x - P_4(x) \right]}{\sqrt{x}}\, dx}_{\equiv G(x)} = \int_{x=0}^{x=1} \left(\frac{\frac{x^5}{5!} + \frac{x^6}{6!} + \cdots}{\sqrt{x}} \right) dx$$

2-Panel Composite Simpson

$$= \frac{.25}{3} \left[0 + 4 \underbrace{(1.70\cdot 10^{-5})}_{\frac{.25^5}{5!\sqrt{.25}} + \frac{.25^6}{6!\sqrt{.25}} + \frac{.25^7}{7!\sqrt{.25}} + \cdots} + 2 \underbrace{(4.013\cdot 10^{-4})}_{\frac{.5^5}{5!\sqrt{.5}} + \frac{.5^6}{6!\sqrt{.5}} + \frac{.5^7}{7!\sqrt{.5}} + \cdots} + 4 \underbrace{(2.6026\cdot 10^{-3})}_{\frac{.75^5}{5!\sqrt{.75}} + \frac{.75^6}{6!\sqrt{.75}} + \frac{.75^7}{7!\sqrt{.75}} + \cdots} + \underbrace{(9.9485\cdot 10^{-3})}_{\frac{1^5}{5!\sqrt{1}} + \frac{1^6}{6!\sqrt{1}} + \frac{1^7}{7!\sqrt{1}} + \cdots} \right]$$

$$= .0017691$$

Sum: $I_1 + I_2 = 2.9235450 + .0017691 = 2.9253141$

The main contribution is computed using the function

$$P_4(x) / x^{1/2} = [1 + x + x^2/2! + x^3/3! + x^4/4!] / x^{1/2} = x^{-1/2} + x^{1/2} + \tfrac{1}{2} x^{3/2} + \tfrac{1}{6} x^{5/2} + \tfrac{1}{24} x^{7/2}$$

and yields a value of 2.9235450. The small correction is obtained by integrating G(x) using a 2-panel Simpson composite method to give .0017691. Note that in computing the first panel of the 2-panel Simpson the fact that the defined function G(x) vanishes at x=0 appears in the first panel explicitly as a "0" for the nodal value.

We can of course improve the result by using a higher degree Taylor polynomial; however, the addition of higher order terms in the truncation error yields smaller and smaller contributions.

It must also be emphasized that this method works only because we know that the integral converges; applying this method to a non-convergent integral produces a result but it will obviously be incorrect.

4.11 Multiple Integrals

Multiple Integrals – 2^d

- General

$$I = \iint_R f(x,y)dxdy = \int_a^b \left\{ \int_c^d f(x,y)dy \right\} dx$$

$$\underbrace{}_{\equiv I_y(x)}$$

$$R = x \in [a,b] \; ; \; y \in [c,d]$$

$$2h = \frac{b-a}{n} \; ; \; 2k = \frac{d-c}{m}$$

- For simplicity take a
  single $2h \times 2k$ cell
  $(m = n = 1)$

$$I_{cell} = \int_{x=x_0}^{x_0+2h} \left\{ \int_{y=y_0}^{y_0+2k} f(x,y)dy \right\} dx$$

$$\underbrace{\phantom{\int_{y=y_0}^{y_0+2k} f(x,y)dy}}_{\equiv I_y(x)}$$

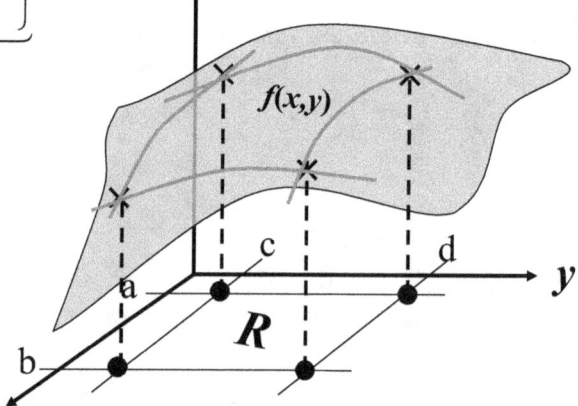

x nodes at x_0, x_0+h, x_0+2h
y nodes at y_0, y_0+k, y_0+2k

Multiple integrals use a 2-dimensional nodal mesh in the x-y plane to specify the z = f(x,y) surface nodal points $f(x_k, y_k)$ as shown in the figure. The nodal spacing along these two axes do not have to be the same and so are generally defined as a uniform spacing of 2h along the x-axis and 2k along the y-axis in anticipation of possible different functional variations along the two axes. Also note that the nodal spacing for the basic cell ("2d panel") has been set to 2h (not h) for notational convenience because we anticipate using Simpson panels, which require three equally spaced points that are normally specified along x as $\{x_0, x_0+h, x_0+2h\}$ and similarly along y. The 2-dimensional region of integration R = {x ε [a,b] , y ε [c,d] } in the x-y plane maps to the patch of surface z = f(x,y) shown in the figure.

For simplicity, we first consider a *single* 2h x 2k cell in the integration region R as a template for the entire integration and then perform the appropriate sum over all cells. All the techniques for single integrals are applicable by first performing the quadrature technique along y to obtain a function of the variable y denoted $I_y(x)$ and then perform the quadrature of $I_y(x)$ along the x coordinates to obtain I_{cell} template shown. (See MatLab script on slide#5-28)

This procedure will define a 2-dimensional pattern of functional evaluations corresponding to a single cell of the specific quadrature and a subsequent summation over all cells yields the 2-dimensional quadrature for the integral I over the region R. The next two slides show the results of this process for Composite Simpson and Gauss quadratures.

4.11.1 Simpson Single 2^d Cell Weight Matrix

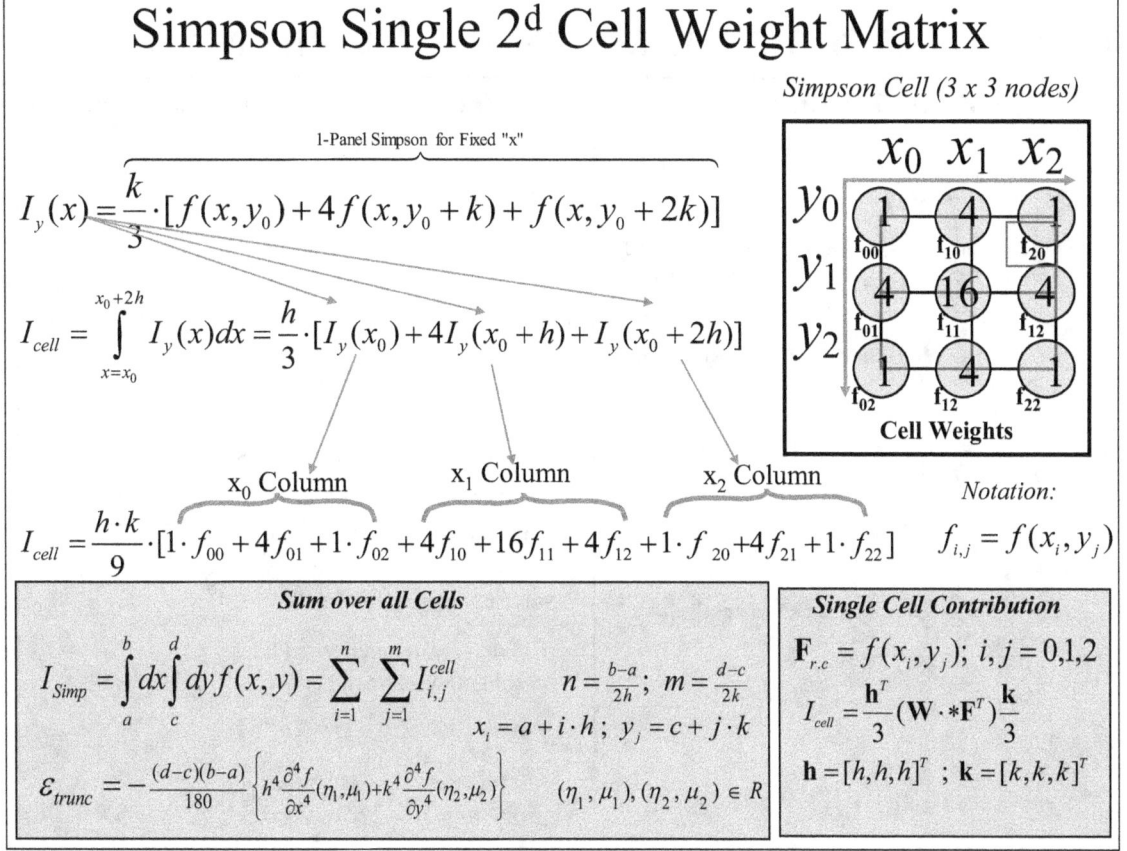

Simpson Single 2^d Cell Weight Matrix

Simpson Cell (3 x 3 nodes)

1-Panel Simpson for Fixed "x"

$$I_y(x) = \frac{k}{3} \cdot [f(x, y_0) + 4f(x, y_0 + k) + f(x, y_0 + 2k)]$$

$$I_{cell} = \int_{x=x_0}^{x_0+2h} I_y(x)\,dx = \frac{h}{3} \cdot [I_y(x_0) + 4I_y(x_0 + h) + I_y(x_0 + 2h)]$$

x_0 Column x_1 Column x_2 Column

Notation:

$$I_{cell} = \frac{h \cdot k}{9} \cdot [1 \cdot f_{00} + 4f_{01} + 1 \cdot f_{02} + 4f_{10} + 16 f_{11} + 4f_{12} + 1 \cdot f_{20} + 4f_{21} + 1 \cdot f_{22}] \qquad f_{i,j} = f(x_i, y_j)$$

Cell Weights (3x3 grid with weights: 1, 4, 1 / 4, 16, 4 / 1, 4, 1)

Sum over all Cells

$$I_{Simp} = \int_a^b dx \int_c^d dy\, f(x,y) = \sum_{i=1}^{n} \sum_{j=1}^{m} I_{i,j}^{cell} \qquad n = \frac{b-a}{2h}; \ m = \frac{d-c}{2k}$$

$$x_i = a + i \cdot h; \ y_j = c + j \cdot k$$

$$\varepsilon_{trunc} = -\frac{(d-c)(b-a)}{180}\left\{ h^4 \frac{\partial^4 f}{\partial x^4}(\eta_1, \mu_1) + k^4 \frac{\partial^4 f}{\partial y^4}(\eta_2, \mu_2) \right\} \qquad (\eta_1, \mu_1), (\eta_2, \mu_2) \in R$$

Single Cell Contribution

$$F_{r.c} = f(x_i, y_j); \ i,j = 0,1,2$$

$$I_{cell} = \frac{\mathbf{h}^T}{3}(\mathbf{W} \cdot \ast \mathbf{F}^T)\frac{\mathbf{k}}{3}$$

$$\mathbf{h} = [h, h, h]^T \ ; \ \mathbf{k} = [k, k, k]^T$$

Applying the general process described on the last slide to the Simpson technique we write down the Simpson panel for $I_y(x)$ given by the first equation. This equation is expressed in terms of the nodal values of the integrand $f(x, y_0)$, $f(x, y_0+h)$, and $f(x, y_0+2h)$ where "x" is of course unspecified.

Next we formally write down an expression for I_{cell} using the Simpson panel for the "new" function $I_y(x)$ given in the second equation. The terms $I_y(x_0)$, $I_y(x_0+h)$, and $I_y(x_0+2h)$ in this (second) equation are now evaluated from the definition of $I_y(x)$ by substituting in turn x= {x_0, x_0+h, x_0+2h} yielding the explicit I_{cell} equation column-by-column. This equation may be interpreted as the fundamental Simpson cell- weighting matrix and is illustrated in the upper right figure. We have used the natural notation that, *e.g.*, f_{12} =f(x_0+1h, x_0+2h), *etc.*, both in the equation and implicitly in the array representation. The transpose $\mathbf{F}_{rc}{}^T$ is used to match the shape of the weight matrix $\mathbf{W}_{rc}$, (which uses y for row and x for column. Because of the symmetry of W, this does not matter for a single cell or for m=n cases, but is needed to match shapes for m ≠ n.

Finally the composite 2^d Simpson quadrature for the integral I_{Simp} over the entire region R is given in the last equation as a sum over all m*n cells. The truncation error is of 4th order in the two stepsizes h along x and k along y and requires 4th partial derivatives with respect to x and y evaluated at two different points in the region R. (See MatLab script on Slides#5-30 to 5-32).

4.11.2   Composite Simpson Multi-Cell Weight Matrix 2^d

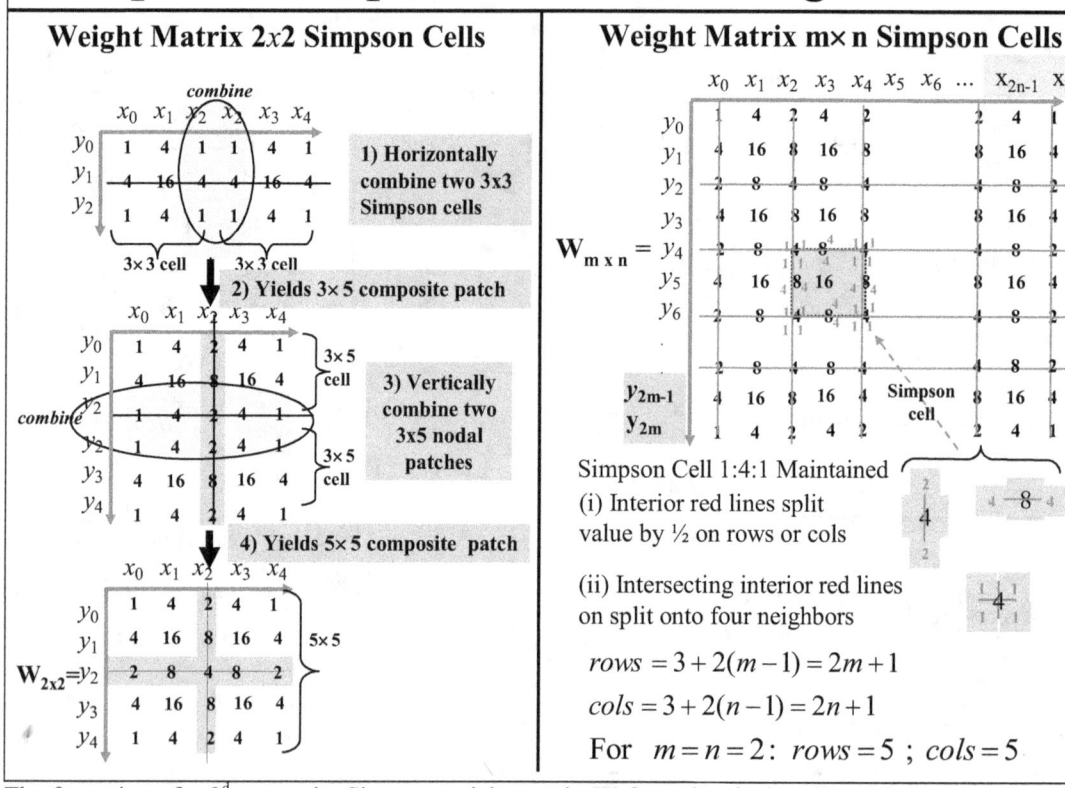

Composite Simpson Multi-Cell Weight Matrix-2^d

The formation of a 2^d composite Simpson weight matrix **W** from the single cell weight matrix shown on the previous slide is illustrated in this slide for the specific case of a 2x2 cell composite with m=n=2. The result is given by the 5 x 5 matrix of weight values which multiply f(x_i, y_j) for i,j = 0 ,1,2,3,4; the weighted sum gives the composite Simpson quadrature shown in the boxed equation of the previous slide. We explicitly show the process for generating the coefficients of a 5 x 5 nodal patch representing a 2 x 2 cell composite Simpson patch in two steps as follows:

1) Two fundamental 3x3 nodal cells are combined horizontally to form a 3 x 5 nodal patch representing a composite 1x2 cell weight matrix (arrow). Note that the common vertical edge terms labeled "x_2" combine to give double their original values thus yielding the 3 new weights for the middle column: W(x_2, y_0)=1+1=2, W(x_2, y_1)=4+4=8, and W(x_2, y_2)=1+1=2 which appear as the vertical shaded column region [2 8 2] in the top right 1x2 cell.

2) This 1x2 cell (3 x 5 nodal patch) is subsequently combined with an identical one below it and similarly the common horizontal edge terms labeled "y_2" to yield (arrow) the double values and the horizontal shaded region [2 8 4 8 2] on the final 2x2 cell weight matrix shown at the bottom of the slide.

The general m x n-cell Simpson composite patch is displayed in the right column above; it simply executes the combining process in the left column m-times horizontally and n-times vertically. For a square matrix illustrated above (m=n=2), the resulting weight matrix **W** has certain obvious symmetries that may be used to combine terms and thereby create a more efficient numerical quadrature algorithm as shown on the next slide (4-37). A general Simpson 2^d MatLab$^©$ algorithm is given in Slides#5-30 to 5-32 for increasing cell structures m=n=10,30,50, giving more accurate results for some examples. The general Simpson 3^d MatLab$^©$ algorithm is also given in Slides#5-33 to 5-36.

4.11.3   Composite 4-Cell Simpson Quadrature Example

Composite 4-Cell Simpson Quadrature Example

Compute a 2 x2 cell composite
Simpson quadrature for the integral

$$I_{exact} = \int_{2.1}^{2.5} dx \int_{1.2}^{1.4} dy\, xy^2 = .3115733$$

Set up the quadrature $n_1 = n_2 = 2$

$h = (b-a)/2n_1 = (2.5-2.1)/2(2) = 0.1$

Change Notation
$(m,n) \rightarrow (n_1,n_2)$

$[a,b] = [2.1,2.5]$

$k = (d-c)/2n_2 = (1.4-1.2)/2(2) = 0.05$

$[c,d] = [1.2,1.4]$

$x_i = a+i\cdot h; \quad y_j = c+j\cdot k$

	2.1 x0	2.2 x1	2.3 x2	2.4 x3	2.5 x4	
y0	3.0240	3.1680	3.3120	3.4560	3.6000	1.20
y1	3.2813	3.4375	3.5937	3.7500	3.9063	1.25
y2	3.5490	3.7180	3.8870	4.0560	4.2250	1.30
y3	3.8272	4.0095	4.1917	4.3740	4.5562	1.35
y4	4.1160	4.3120	4.5080	4.7040	4.9000	1.40

$\mathbf{F^T} = $ (left table)

	2.1 x0	2.2 x1	2.3 x2	2.4 x3	2.5 x4
y0	1	4	2	4	1
y1	4	16	8	16	4
y2	2	8	4	8	2
y3	4	16	8	16	4
y4	1	4	2	4	1

$\mathbf{W} = $ (right table)

$$I_{Simp} = \frac{h\cdot k}{9}\cdot\sum_{i=0}^{4}\sum_{j=0}^{4}\left[\mathbf{W}.*\mathbf{F}^T\right] = \frac{(0.1)(0.05)}{9}\cdot\sum_{i=0}^{4}\sum_{j=0}^{4}\left[\mathbf{W}.*\mathbf{F}^T\right] = .3115733$$

$$\varepsilon_{trunc} = -\frac{(d-c)(b-a)}{180}\left\{h^4\frac{\partial^4 f}{\partial x^4}(\eta_1,\mu_1)+k^4\frac{\partial^4 f}{\partial y^4}(\eta_2,\mu_2)\right\} = -\frac{(.4)(.2)}{180}\left\{(.1)^4\cdot 0+(.05)^4\cdot 0\right\} = 0$$

4th partial derivatives are both zero

Here is a simple example of a 2-dimensional composite Simpson quadrature for the integral shown at the top of the slide. Setting up the quadrature is quite straightforward for this case $n_1=n_2=2$ as we have already computed the 5x5 composite weighting matrix $\mathbf{W}$ on the previous slide (repeated here for convenience). The values of the parameters [a,b], [c,d] are computed directly from the integral limits, and the stepsizes h and k are computed to be 0.1 and 0.05 respectively. The Simpson quadrature is obtained by taking the point-by-point (MatLab ".*") matrix product and then summing all terms in the resulting product matrix. The result is seen to agree precisely with the exact integral; this precise agreement results from the vanishing of the two fourth partial derivatives which gives zero truncation error. This *will not generally be the case* with non-polynomial integrands or with polynomials of degree>3 in either x or y; in these cases one or both of the 4th order partials f_{xxxx}, f_{yyyy} will contribute to the truncation error. In Slide# 5-28, we give a short MatLab© script to compute 4-cell composite Simpson double integral and a 3-pt Gaussian Quadrature discussed on the last slide of this section. The next few slides give a general formulation of Simpson 2-dim and 3-dim integrals for non-regular cell curves c(x) and d(x) and surfaces alp(x,y) and beta(x,y). (See MatLab© quadrature scripts on Slides# 5-33 to 5-39)

4.11.4 Simpson Non-Rectangular Cell – 2^d

Simpson Non-Rectangular Cell - 2^d

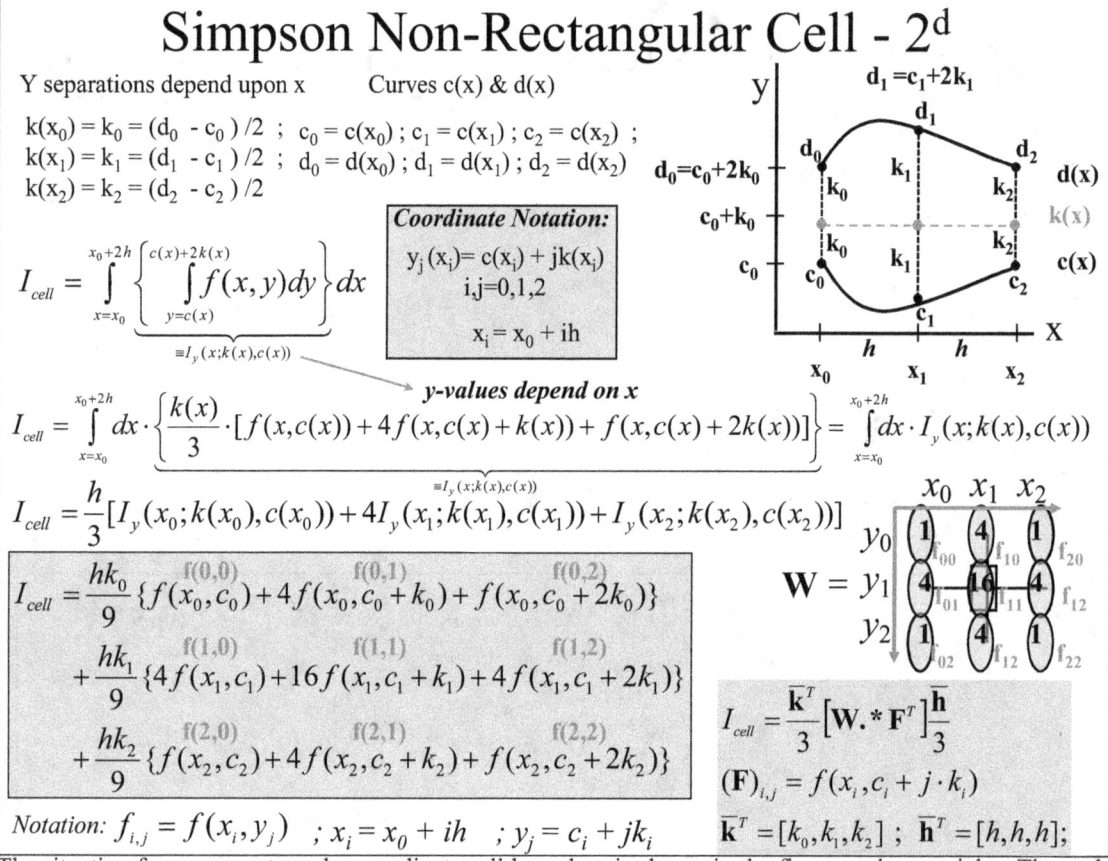

Y separations depend upon x   Curves c(x) & d(x)

$k(x_0) = k_0 = (d_0 - c_0)/2$; $c_0 = c(x_0)$; $c_1 = c(x_1)$; $c_2 = c(x_2)$;
$k(x_1) = k_1 = (d_1 - c_1)/2$; $d_0 = d(x_0)$; $d_1 = d(x_1)$; $d_2 = d(x_2)$
$k(x_2) = k_2 = (d_2 - c_2)/2$

Coordinate Notation:
$y_j(x_i) = c(x_i) + jk(x_i)$
$i,j = 0,1,2$
$x_i = x_0 + ih$

$$I_{cell} = \int_{x=x_0}^{x_0+2h} \underbrace{\left\{ \int_{y=c(x)}^{c(x)+2k(x)} f(x,y)dy \right\}}_{\equiv I_y(x;k(x),c(x))} dx$$

y-values depend on x

$$I_{cell} = \int_{x=x_0}^{x_0+2h} dx \cdot \underbrace{\left\{ \frac{k(x)}{3} \cdot [f(x,c(x)) + 4f(x,c(x)+k(x)) + f(x,c(x)+2k(x))] \right\}}_{\equiv I_y(x;k(x),c(x))} = \int_{x=x_0}^{x_0+2h} dx \cdot I_y(x;k(x),c(x))$$

$$I_{cell} = \frac{h}{3}[I_y(x_0;k(x_0),c(x_0)) + 4I_y(x_1;k(x_1),c(x_1)) + I_y(x_2;k(x_2),c(x_2))]$$

$$I_{cell} = \frac{hk_0}{9}\{f(x_0,c_0) + 4f(x_0,c_0+k_0) + f(x_0,c_0+2k_0)\}$$
$$+ \frac{hk_1}{9}\{4f(x_1,c_1) + 16f(x_1,c_1+k_1) + 4f(x_1,c_1+2k_1)\}$$
$$+ \frac{hk_2}{9}\{f(x_2,c_2) + 4f(x_2,c_2+k_2) + f(x_2,c_2+2k_2)\}$$

f(0,0) f(0,1) f(0,2)
f(1,0) f(1,1) f(1,2)
f(2,0) f(2,1) f(2,2)

$$\mathbf{W} = \begin{array}{c} y_0 \\ y_1 \\ y_2 \end{array}\begin{array}{ccc} x_0 & x_1 & x_2 \\ 1 & 4 & 1 \\ 4 & 16 & 4 \\ 1 & 4 & 1 \end{array}$$

$$I_{cell} = \frac{\overline{k}^T}{3}\left[\mathbf{W}.*\mathbf{F}^T\right]\frac{\overline{h}}{3}$$
$$(\mathbf{F})_{i,j} = f(x_i, c_i + j \cdot k_i)$$

Notation: $f_{i,j} = f(x_i, y_j)$; $x_i = x_0 + ih$; $y_j = c_i + jk_i$ $\overline{k}^T = [k_0,k_1,k_2]$; $\overline{h}^T = [h,h,h]$;

The situation for a non-rectangular coordinate cell boundary is shown in the figure at the top right. The nodes along the x-axis are taken to be x_0, $x_1 = x_0 + h$, $x_2 = x_0 + 2h$, while those along the y axis change as a function of x. The two boundary curves in the figure c(x) and d(x) take on the y-values along each curve $c_0 = c(x_0)$, $c_1 = c(x_1)$, $c_2 = c(x_2)$ and $d_0 = d(x_0)$, $d_1 = d(x_1)$, $d_2 = d(x_2)$. In order to define equally spaced y-nodes for the y-integration at *each* x-value, we must compute the intermediate points $k_0 = (d_0 - c_0)/2$, $k_1 = (d_1 - c_1)/2$, $k_2 = (d_2 - c_2)/2$. Inspection of the figure shows that the y-values at x_0 may be written in the form $\{c_0, c_0+k_0, c_0+2k_0 = d_0\}$; similarly at x_1 they are $\{c_1, c_1+k_1, c_1+2k_1 = d_1\}$ and at x_2 they are $\{c_2, c_2+k_2, c_2+2k_2 = d_2\}$.

Given these definitions, the double integral is developed in a manner similar to that for the rectangular case. As shown on the slide, the double integral is first integrated over y to yield a new function of x:

$$I_y(x; c(x), k(x)) = (k(x)/3)[f(x, c(x)) + 4 f(x, c(x)+k(x)) + f(x, c(x)+2k(x))]$$

Next the x integration is written down in the usual 1-dimensional fashion in terms of this new function

$$I_{cell} = (h/3)[I_y(x_0; c(x_0), k(x_0)) + 4 I_y(x_1; c(x_1), k(x_1)) + I_y(x_2; c(x_2), k(x_2))]$$

Finally the explicit expression for each $I_y(x; c(x), k(x))$ is substituted and then expanded out to yield the 9 terms shown in the boxed equation at the bottom. The weighting matrix **W** is the same as for the rectangular case; however, there are different k(x) values for each row of the expression. If we use the shorthand notation f(i, j) shown above each term in the boxed equation and further define the vectors $\mathbf{k} = [k_0, k_1, k_2]^T$, $\mathbf{h} = [h, h, h]^T$, and the matrix $(\mathbf{F_k})_{ij} = f(x_i, c_i + j \cdot k_i)$, then the expression for the integral yields a quadratic form $I_{cell} = \mathbf{k}^T/3\ [\mathbf{W}.*\mathbf{F}^T]$ $\mathbf{h}/3$, where ".*" represents element-by-element matrix multiplication as in MatLab©. Again the transpose $\mathbf{F}^T$ is used the shape of **W** for multiple cells.

4.11.5 Simpson Non-Rectangular Cell – 3^d

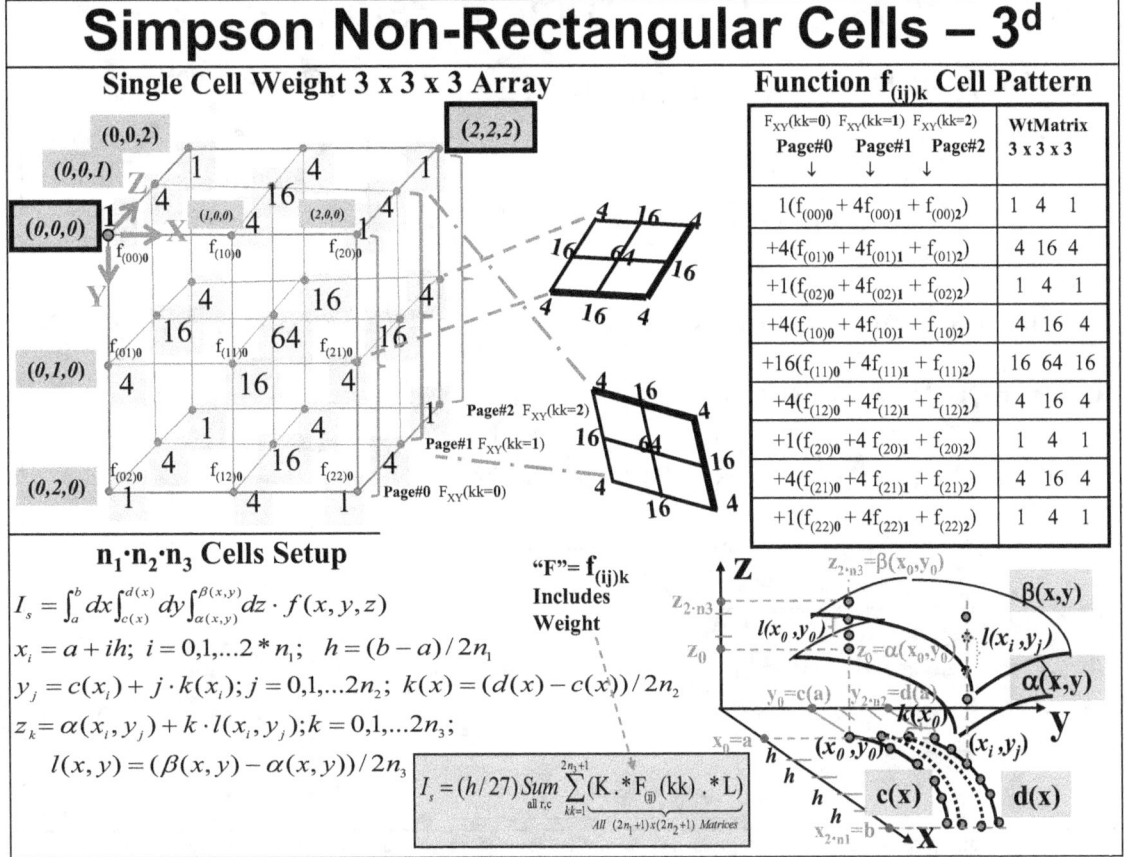

Function $f_{(ij)k}$ Cell Pattern

$F_{XY}(kk=0)$ Page#0 $\downarrow$	$F_{XY}(kk=1)$ Page#1 $\downarrow$	$F_{XY}(kk=2)$ Page#2 $\downarrow$	WtMatrix 3 x 3 x 3
$1(f_{(00)0} + 4f_{(00)1} + f_{(00)2})$			1 4 1
$+4(f_{(01)0} + 4f_{(01)1} + f_{(01)2})$			4 16 4
$+1(f_{(02)0} + 4f_{(02)1} + f_{(02)2})$			1 4 1
$+4(f_{(10)0} + 4f_{(10)1} + f_{(10)2})$			4 16 4
$+16(f_{(11)0} + 4f_{(11)1} + f_{(11)2})$			16 64 16
$+4(f_{(12)0} + 4f_{(12)1} + f_{(12)2})$			4 16 4
$+1(f_{(20)0} +4 f_{(20)1} + f_{(20)2})$			1 4 1
$+4(f_{(21)0} +4 f_{(21)1} + f_{(21)2})$			4 16 4
$+1(f_{(22)0} + 4f_{(22)1} + f_{(22)2})$			1 4 1

An expansion, similar to that for double integral, starts at the inner z-coordinate and then back through y and then x coordinates to yield the cell structure for the function f(x,y,z) and the 3^d cell weight matrix given on this slide. In extending to n_1, n_2, n_3 cells in the x, y, and z directions respectively, the assignment of rows, columns, and pages for the **composite weight array** differs from that of the **natural function array** f(x,y,z) Thus for multiple cells the key requirement is that the shapes of these two arrays have matching dimensions, *e.g.*, for the # cells = $[n_1,n_2,n_3]$ = [1,2,3], both arrays must have [row, col, page] = $[(2n_1+1),(2n_2+1),(2n_3+1)]$ = [3, 5, 7] and *only in that order*.

The function array pattern $F_{(ij)}(kk)$ is naturally written as a $(2n_1+1) \times (2n_2+1)$ matrix array with index kk running through the $(2n_3+1)$ pages. It includes the weight array $W_{(ij)}(kk)$ in its definition as given in the table. The Simpson composite integral given in the boxed equation is arrived at (see bottom figure) as follows:
(i) The x-values are incremented by a constant scalar value h which yields the multiplier h/27 outside all sums;
(ii) The y-values, $k(x)$ depend upon the y-distance between the curves c(x) and d(x) and are therefore different for each x value. These values are re-expressed as a $(2n_1+1) \times (2n_2+1)$ matrix **K** for all x and y, and then are "point multiplied" with $F_{(ij)}(kk)$ to give K .* $F_{(ij)}(kk)$;
(iii) The z-values, $l(x,y)$ depend upon the z-distance between the surfaces $\alpha(x,y)$ and $\beta(x,y)$ and are therefore different for each (x,y) pair value. These values are re-expressed as a $(2n_1+1) \times (2n_2+1)$ matrix **L** and again "point multiplied" to yield (K .* $F_{(ij)}(kk)$.* L.
(iv) The final result is obtained by first summing the matrices over all pages (index *kk*) and then taking row and column sums of the resulting matrix, $(2n_1+1) \times (2n_2+1)$ matrix which yields the scalar value of the integral, *viz.*,

$$I_s = (h/27)\,Sum_{\text{all r,c}} \sum_{kk=1}^{2n_3+1}(K .* F_{(ij)}(kk) .* L)$$

$$\underbrace{}_{\text{All } (2n_1+1)x(2n_2+1) \text{ Matrices}}$$

The slide illustrates the structure in detail and the resulting 2^d and 3^d MatLab$^©$ scripts are given in the appendix on Slides #5-33 to 5-39.

Gauss Cell Coefficient Pattern - 2^d

Gaussian Quadrature

$$I_{exact} = \int_{2.1}^{2.5} dx \int_{1.2}^{1.4} dy\, xy^2 = .3115733$$

Mapping

$R: x \in [a,b] = [2.1,2.5] \; ; \; y \in [c,d] = [1.2,1.4]$

$x = \dfrac{a+b}{2} + \dfrac{b-a}{2} u = 2.3 + .2u \; ; \; y = \dfrac{c+d}{2} + \dfrac{d-c}{2} v = 1.3 + .1v$

$dx = .2du \; ; \; dy = .1du \qquad \Rightarrow \tilde{R} = u \in [-1,+1] \; ; \; v \in [-1,+1]$

$$I_G = \int_{-1}^{1} du \int_{-1}^{1} dv\, (.02)(2.3 + .2u_i)(1.3 + .1v_j)^2 \quad \therefore g(u_i, v_j) = (.02)(2.3 + .2u_i)(1.3 + .1v_j)^2$$

Zeros & Coeffs. (n=3)

$u_0, v_0 = .77459 \; ; \; c_0 = .55555$

$u_1, v_1 = .00000 \; ; \; c_1 = .88889$

$u_2, v_2 = -.77459 \; ; \; c_2 = .55555$

Gaussian Quadrature Patch (3 pt x 3pt)

$$I = \int_a^b \int_c^d f(x,y)\,dxdy = \int_{-1}^{+1}\left\{\int_{-1}^{+1} g(u,v)\,du\right\}dv$$

$$= \sum_{j=0}^{2} c_j \left(\underbrace{\sum_{i=0}^{2} c_i g(u_i, v_j)}_{= \sum_{i=0}^{2} c_i g(u_i, v_j)} \right)$$

$$I = \sum_{i=0}^{2}\sum_{j=0}^{2} c_i c_j g(u_i, v_j)$$

$$g(u_i, v_j) \equiv \left(\frac{b-a}{2}\right)\cdot\left(\frac{d-c}{2}\right)\cdot f(\frac{a+b}{2} + \frac{b-a}{2}u \; , \; \frac{c+d}{2} + \frac{d-c}{2}v)$$

or

$$I = \sum_{i=0}^{2}\sum_{j=0}^{2} C.*G$$

$$G \equiv g(u_i, v_j)$$

$$C \equiv c_i c_j$$

The Gaussian quadrature for multiple integrals uses a similar procedure to that for Simpson except for a number of notational conventions more suited to Gaussian quadrature. As in the Simpson case, we set up the quadrature for a single cell and then sum over all cells in the region R. Alternately the Gauss quadrature may be applied to the region R as a whole by adding more points to give a Gauss Quadrature based on a higher degree Legendre function.

The transformation of the region R = {x ε [a,b] , y ε [c,d] } to R_G = {u ε [-1.1] , v ε [-1,1] } requires the two transformations shown in the first group of equations and results in a new function g(u,v) given in the boxed equation at the bottom of the slide. The zeros and coefficients given in the slide are identical in each direction assuming of course that we use the same degree Legendre polynomials for both u and v.

For a 3-point Gaussian quadrature we follow the same procedure as with Simpson, first writing down the Gaussian quadrature along the u coordinate to obtain $I_u(v)$ as

$$I_u(v) = c_0\, g(u_0,v) + c_1\, g(u_0+h,v) + c_2\, g(u_0+2h, v),$$

and then performing a Gaussian quadrature along the v coordinate to obtain I

$$I = c_0\, I_u(v_0) + c_1\, I_u(v_1) + c_2\, I_u(v_2)$$

Substitution of $I_u(v)$ for each value of v yields the double sum given in the boxed equation; the function g(u,v) expressed in terms of the original parameters a,b,c,d and f(x,y) is also given in the boxed equation. This process is easily generalized to n-point Gauss-Legendre quadratures as well as Gauss-Tschebychev, Gauss-Hermite, and Gauss-Laguerre. Note that by defining the matrices C and G as in the slide, the result is again expressed as a double sum over a element-by-element MatLab[©] matrix product. A short MatLab[©] script (Slide# 5-28) is applied to the same double integral evaluated on Slide #4-37 by Simpson Quadrature. The Gaussian quadrature requires only 9 functional evaluations of $g(u_i, v_j)$ as compared to 25 evaluations of $f(x_i, y_j)$ for the Simpson method. Also see general 2^d Gaussian MatLab[©] scripts on Slides #5-30 to 5-32.

The general formulation of 2^d and 3^d Gaussian quadratures with non-rectangular cells is determined by the geometry shown for Simpson derivation of the last slide and in general requires different transformations for each of the non-rectangular y and z spans, which adds some complexity to the Gaussian computations. A brief summary of the formulation of the 3^d Gaussian triple integral is given in the next slide.

Gaussian Non-Rectangular Cells – 3^d

Transformation to [-1,1]

$$I_G = \int_a^b dx \int_{c(x)}^{d(x)} dy \int_{\alpha(x,y)}^{\beta(x,y)} dz \cdot f(x,y,z)$$

$$x_i = h_1 \cdot u + h_2 \; ; \quad h_1 = (b-a)/2; \; h_2 = (b+a)/2$$

$$y_j = k_1 \cdot v + k_2 \; ; \quad k_1 = (d(x) - c(x))/2; \; k_2 = (d(x) + c(x))/2$$

$$z_k = l_1 \cdot w + l_2 \; ; \quad l_1 = (\beta(x,y) - \alpha(x,y))/2; \; l_2 = (\beta(x,y) + \alpha(x,y))/2$$

$$I_G = \int_{-1}^{+1} du \int_{-1}^{+1} dv \int_{-1}^{+1} dw \cdot g(u,v,w)$$

$$g(u,v,w) = h_1 \cdot k_1(u) \cdot l_1(u,v) \cdot f(h_1 \cdot u + h_2, \; k_1(u) \cdot v + k_2(u), \; l_1(u,v) \cdot w + l_2(u,v))$$

Gaussian Roots and Coefficients

$$u_i = rn_1(i); c_i = cn_1(i);$$
$$v_j = rn_2(j); c_j = cn_2(j);$$
$$w_k = rn_3(k); c_k = cn_3(k);$$

$$I^G_w = \int_{-1}^{+1} du \int_{-1}^{+1} dv \sum_{kk=1}^{n_3} cn_3(k) \cdot g(u,v,rn_3(k))$$

$$I^G_{vw} = \int_{-1}^{+1} du \sum_{j=1}^{n_2} cn_2(j) \sum_{kk=1}^{n_3} cn_3(k) \cdot g(u,rn_2(j),rn_3(k))$$

$$I^G_{uvw} = \sum_{i=1}^{n_1} cn_1(i) \sum_{j=1}^{n_2} cn_2(j) \sum_{kk=1}^{n_3} cn_3(k) \cdot g(rn_1(i),rn_2(j),rn_3(k))$$

$$I^G_{uvw} = \sum_{kk=1}^{n_3} cn_3(k) \sum_{i=1}^{n_1} \sum_{j=1}^{n_2} cn_1(i) \cdot cn_2(j) \cdot g(rn_1(i),rn_2(j),rn_3(k))$$

$$\boxed{I^G = Sum_{Page}(Sum_{Row}(Sum_{Col}(C_{ij,k} \cdot {}^* G_{ij,k})))}$$

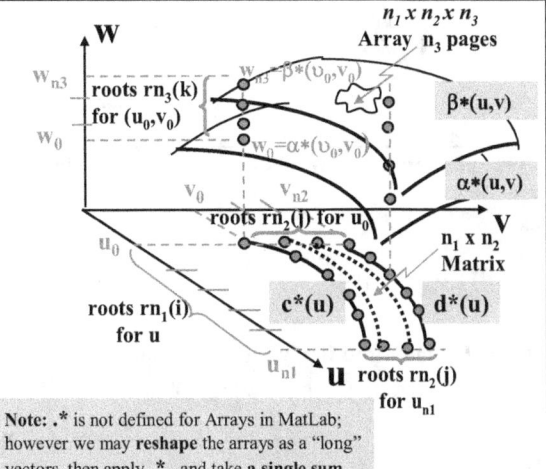

Note: .* is not defined for Arrays in MatLab; however we may **reshape** the arrays as a "long" vectors, then apply .* , and take **a single sum**

The *upper panel* describes the transformation of the original triple integral of the function f(x,y,z) with integration ranges x ε [a,b], y ε [c(x), d(x)], and z ε [α(x,y), β(x,y)] to the Gaussian Quadrature form integrating a new function g(u,v,w), with integration ranges {u, v, w} ε [-1, 1], is given by

$$g(u,v,w) = h_1 \cdot k_1(u) \cdot l_1(u,v) \cdot f(h_1 \cdot u + h_2, \; k_1(u) \cdot v + k_2(u), \; l_1(u,v) \cdot w + l_2(u,v))$$

where the transformation parameters {h₁, h₂, k₁(u), k₂(u), l₁(u,v), l₂(u,v)} and implicit expressions for {x(u), y(u,v), z(u,v,w)} are defined in the upper panel in terms of the original integration limits, [a,b], [c(x), d(x)], [α(x,y),β(x,y)].

The *lower panel* equations define the Legendre polynomial roots rn₁(i), rn₂(j), rn₃(k) and coefficients cn₁(i), cn₂(j), cn₃(k) for each dimension and the steps extend the 2^d derivation to yield the quadrature

$$I^G = Sum_{Page}(Sum_{Row}(Sum_{Col}(C_{ij,k} \cdot {}^* G_{ij,k})))$$

Note: Pt-by-Pt .* is not defined for Arrays in MatLab; however we may **reshape** the arrays as "long" vectors, then apply .* , and take **a single sum** to yield

$$I^G = Sum(long\_C_{ij,k} \cdot {}^*long\_G_{ij,k})$$

The figure in the bottom panel illustrates the geometry in u,v,w-coordinates showing Legendre roots along u, and roots along v illustrated for two fixed values ,i.e., u=u₀ and u=u_{n1}; this process forms a **n₁ x n₂ matrix** as indicated. The roots along w illustrated at one point (u₀, v₀) fill out the points between the two surfaces α(u,v) and β(u,v); this process across all (u,v) points yields the **n₁ x n₂ x n₃ Array** shown. The MatLab© scripts are given in the appendix Slides# 5-33 to 5-35.

4.11.8   Simpson Gauss Quadratures Comparison - Cone Example

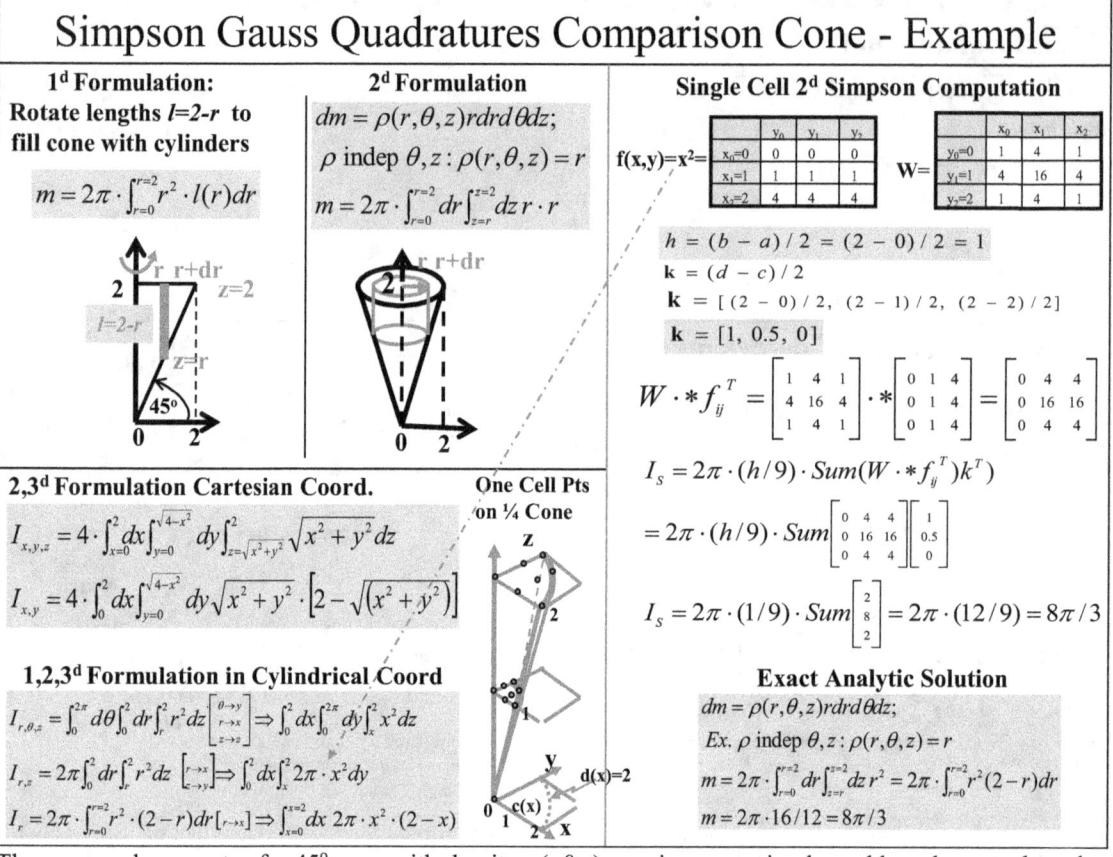

Simpson Gauss Quadratures Comparison Cone - Example

1^d Formulation:
Rotate lengths l=2-r to fill cone with cylinders

$$m = 2\pi \cdot \int_{r=0}^{r=2} r^2 \cdot l(r)dr$$

2^d Formulation

$$dm = \rho(r,\theta,z)rdrd\theta dz;$$

ρ indep $\theta,z : \rho(r,\theta,z) = r$

$$m = 2\pi \cdot \int_{r=0}^{r=2} dr \int_{z=r}^{z=2} dz\, r \cdot r$$

Single Cell 2^d Simpson Computation

$$f(x,y)=x^2=$$

	y_0	y_1	y_2
x_0=0	0	0	0
x_1=1	1	1	1
x_2=2	4	4	4

$$W=$$

	x_0	x_1	x_2
y_0=0	1	4	1
y_1=1	4	16	4
y_2=2	1	4	1

$h = (b-a)/2 = (2-0)/2 = 1$

$k = (d-c)/2$

$k = [(2-0)/2, (2-1)/2, (2-2)/2]$

$k = [1, 0.5, 0]$

$$W \cdot * f_{ij}^T = \begin{bmatrix} 1 & 4 & 1 \\ 4 & 16 & 4 \\ 1 & 4 & 1 \end{bmatrix} \cdot * \begin{bmatrix} 0 & 1 & 4 \\ 0 & 1 & 4 \\ 0 & 1 & 4 \end{bmatrix} = \begin{bmatrix} 0 & 4 & 4 \\ 0 & 16 & 16 \\ 0 & 4 & 4 \end{bmatrix}$$

$$I_S = 2\pi \cdot (h/9) \cdot Sum(W \cdot * f_{ij}^T)k^T)$$

$$= 2\pi \cdot (h/9) \cdot Sum \begin{bmatrix} 0 & 4 & 4 \\ 0 & 16 & 16 \\ 0 & 4 & 4 \end{bmatrix} \begin{bmatrix} 1 \\ 0.5 \\ 0 \end{bmatrix}$$

$$I_S = 2\pi \cdot (1/9) \cdot Sum \begin{bmatrix} 2 \\ 8 \\ 2 \end{bmatrix} = 2\pi \cdot (12/9) = 8\pi/3$$

2,3^d Formulation Cartesian Coord.

$$I_{x,y,z} = 4 \cdot \int_{x=0}^{2} dx \int_{y=0}^{\sqrt{4-x^2}} dy \int_{z=\sqrt{x^2+y^2}}^{2} \sqrt{x^2+y^2}\, dz$$

$$I_{x,y} = 4 \cdot \int_{0}^{2} dx \int_{y=0}^{\sqrt{4-x^2}} dy \sqrt{x^2+y^2} \cdot \left[2 - \sqrt{(x^2+y^2)}\right]$$

One Cell Pts on ¼ Cone

1,2,3^d Formulation in Cylindrical Coord

$$I_{r,\theta,z} = \int_{0}^{2\pi} d\theta \int_{0}^{2} dr \int_{r}^{2} r^2 dz \begin{bmatrix} \theta \to y \\ r \to x \\ z \to z \end{bmatrix} \Rightarrow \int_{0}^{2} dx \int_{0}^{2\pi} dy \int_{x}^{2} x^2 dz$$

$$I_{r,z} = 2\pi \int_{0}^{2} dr \int_{r}^{2} r^2 dz \begin{bmatrix} r \to x \\ z \to y \end{bmatrix} \Rightarrow \int_{0}^{2} dx \int_{x}^{2} 2\pi \cdot x^2 dy$$

$$I_r = 2\pi \cdot \int_{r=0}^{r=2} r^2 \cdot (2-r)dr [r \to x] \Rightarrow \int_{x=0}^{x=2} dx\, 2\pi \cdot x^2 \cdot (2-x)$$

Exact Analytic Solution

$$dm = \rho(r,\theta,z)rdrd\theta dz;$$

Ex. ρ indep $\theta, z : \rho(r,\theta,z) = r$

$$m = 2\pi \cdot \int_{r=0}^{r=2} dr \int_{z=r}^{z=2} dz\, r^2 = 2\pi \cdot \int_{r=0}^{r=2} r^2(2-r)dr$$

$$m = 2\pi \cdot 16/12 = 8\pi/3$$

The mass and moments of a 45° cone with density $\rho(r,\theta,z) = r$ is a very simple problem that may be solved analytically and can be set up as either a 1-, 2-, or 3-dimensional integral. It is therefore a useful case in understanding the general Simpson and Gauss quadrature methods in Cartesian x,y,z coordinates. The simplest way to calculate the mass of the axially distributed density, $\rho(r\ y\ z) = r$ is to generate the cone by rotating the triangle by 2π around the z-axis. The red vertical line segment (*length l = 2-r*) then yields constant density cylinder contributions to the mass and results in a 1^d integral over the axial distance r ε [0,2] from r=0 to r=2 . Both single cell Simpson and 2-pt Gaussian Gaussian Quadrature yield full numerical precision as both are precise up to *cubic polynomials* and the integrand $2\pi\ r^2(2-r)$ is a cubic polynomial in r. Similarly, the 2^d integral over r ε [0,2] and z ε [r, 2] yields the same cubic integrand $2\pi\ r^2(2-r)$ and thus again both single cell Simpson and n=1 Gaussian Quadrature yield exact results.

However, for the 3^d integral over x ε [0, 2] and y ε [0, sqrt(4-x²)], z ε [sqrt(4-x²), 2], the term in the integrand sqrt(x²+y²) is not a polynomial, so truncation errors result. Clearly, the 3^d integral set up in (x,y,z) coordinates is more complex then it needs to be considering the rotational symmetry about the z-axis; the cylindrical coordinate box (bottom left) shows all three transformed integrals are at most cubic polynomials and so exact results. It is far better to set the integral in appropriate 3^d cylindrical coordinates (r, θ ,z) which is immediately re-written as a 2^d integral over r and z. The comparison table on the next slide illustrates these results for 1^d, 2^d, & 3^d integrals using Composite Simpson for a number of composite cases with n-cells (taking n =n_1=n_2=n_3) and for a number of n-pt Gaussian Quadratures.

4.11.9 Multiple Integrals Methods Comparison

Multiple Integration Methods Comparison

Simpson Composite *vs.* Gaussian Quadrature 2^d

Ex 1 fij=exp(yj/xi); a=0.1; b=0.5; c=xi³; d=xi²;

n	I_Simpson	I_Gauss	n
	*	0.0333453874623914	2
	*	0.0333058313348074	3
		0.0333055670528154	4
2	0.03249434240621	*	
3	0.03330476252307	*	
5	0.03330546128190	0.0333055661186752	5
10	0.03330555954577	*	
30	0.03330556603505	0.0333055661145462	10

I_exact = 0.03330556611000

Full Precision

Ex 2 fij=(yj³ + xi²); a=2; b=2.2; c=xi; d=2*xi;

n	I_Simpson	I_Gauss	n
1	16.50865000000002	*	
2	16.24616523437502	16.50864	3
4	16.50864003906252	16.50864	4
5	16.50864001600002	16.5086399999999	5
8	16.50864000244143	*	
16	16.50864000015261	*	
50	16.50864000000162	16.5086399990399	10

I_exact =16.50864000000

Ex 3 fij=xi; a=0; b=1; c=xi; d=1

n	I_Simpson	I_Gauss	n
1	0.16666666666667	*	
2	0.16666666666667	0.166666666666667	2
3	0.16666666666667	0.166666666666667	3

I_exact =1/6 = 0.166666666666667

Full Precision

2^d Quadrature Cylindrical Coord.

Ex 4 fij=xi; a=0; b=2; c=2-xi; d=2;45°Cone Cyl.Coord.

n	I_Simpson	I_Gauss	n
1	8.37758040957278	8.37758040957278	2
2	8.37758040957278	8.37758040957278	3

I_exact = 8.37758040957278

Full Precision

3^d Reduced to 2^d Cartesian *cf.* 3^d Gaussian Quadrature

Ex 5 fij=4*sqrt(xi²+yj²)*(2-sqrt(xi²+yj²)); a=0; b=2; c=0; d=sqrt(4-x²); Cone

#	3^d → 2^d	3^d
2	8.01804432288262	8.44265651359811
3	8.34054363148634	8.39837596631408
5	8.3750447446989	8.38259185971219
7	8.37703491194628	*
10	8.37751123138186	8.378332223375277

I$_{exact}$ =8*pi/3 = 8.37758040957

The comparison tables on this slide are a sampling of tables generated by the MatLab$^©$ scripts given in the appendix (slides #5-30 to 5-36). In particular, we have chosen tables that represent the various approaches that can be used to solve the 45°-cone problem that we just discussed on the previous Slide#4-42. Each table gives a comparison of 2^d Composite Simpson with *n-cells* (taking n =n$_1$=n$_2$) and an *n-point* Gaussian Quadrature. The solution to the cone problem with axial density ρ(r, θ, z) = r, yields exact numerical precision for n$_{Simp}$ =1 as well as for n$_{Gauss}$=2 pts, and does not change as more cells are added (see **Table for Ex. # 4**)

This is not unexpected because the integrated polynomial in each case does not exceed cubic and both methods have cubic polynomial precision. For Simpson, the same cubic precision remains in effect, even as the number of cells n$_{Simp}$ increase, whereas, Gaussian Quadrature generates polynomial precision (2 n$_{Gauss}$-1) which increases as points are added. Comparing the two columns shows that 2^d integration has somewhat faster approach to the exact value, so it always pays to do the analytic pre-integration to reduce the dimension when possible.

Ex#5 compares a 3^d Gaussian Quadrature to one that is reduced to 2^d *via* an analytic z-integration yielding a new 2^d function f(x,y) = $4\sqrt{x^2+y^2} \cdot \left[2-\sqrt{(x^2+y^2)}\right]$ (as discussed on the previous slide). This should be compared with the full precision 2^d Cylindrical Coordinate Gauss or Simpson Quadratures of **Ex. # 4**.

The following statements appear to be the case: (i)1^d Simpson and Gaussian Quadrature work well and are fast; (ii)2^d Simpson memory and time could be issues for n$_{Simp}$ > 256; (iii)3^d Simpson memory and time are issues for n > 64.

In general, Simpson can be used with *adaptive stepsize* methods, whereas Gaussian Quadrature stepsize is not adjustable inasmuch as the fixed roots are pre-determined by roots to Legendre polynomials. However, this assignment yields higher polynomial precision with increasing number of points. To fit multinomials in x, y, z, the order of the Legendre polynomials must be adapted to each dimension for best precision. Finally, the method of stating problem can be crucial and choosing coordinates adapted to the specific symmetry improves efficiency of the Quadrature.

5 MatLab Scripts

1. Derivative of exp(x)
2. Polynomial Cancellation
3. Fixed Point Example
4. Aitken's Convergence Acceleration
5. Taylor Polynomials for 1/x
6. Lagrange Polynomials for Sin(x) and Asin(x)
7. Richardson Extrapolation
8. Bezier Curves
9. Bezier Curves – Interactive
10. Root Finding Methods Comparison
11. Simpson 1^d Composite with fixed intervals
12. Simpson 1^d Adaptive Quadrature
13. Gauss 1^d Multi-Panel Quadrature
14. Quadratures – Simple 2^d Simpson & Gaussian
15. General 2^d Simpson Quadrature & Weight Matrix
16. General 3^d Simpson Quadrature & Weight Matrix
17. General 2^d Gaussian Quadrature
18. General 3^d Gaussian Quadrature

5.1 *Derivative of exp(x)*

```
% Numerical Derivative is Zero Demo
clear; d=5; format long g;
digits(d);                          % set computer wordsize "digits"  d=5
f1=inline('exp(x)');                 %Define exponential function
k=1:20;                              % vector to "loop over"
hk=2.^-k;                           % "h-value": 2 raised to the "vector" -k
hk=vpa(hk);                          %symbolic round off to d digits
Dk=zeros(length(k),1);
Dk=vpa((vpa(f1(1+hk))-vpa(f1(1))). /hk);
delta=abs(vpa(Dk)-vpa(f1(1)));
Dk=double(Dk);
delta=double(delta);
hk=double(hk);
{'1st Deriv'  'hk' 'error'}
[Dk',hk', delta' ]
figure(1), hold on
plot(-log2(hk),Dk,'*-')
plot(-log2(hk), repmat(exp(1),1,length(hk)),'r:','linewidth',2)
%plot(-log2(hk), log10(delta),'g:'), hold off
```

5.2 *Polynomial Cancellation*

```
%Polynomial Cancellation "polycancel.m"
close all
f1=inline('(x-1).^6');
f2=inline('x.^6-6*x.^5+15*x.^4-20*x.^3+15*x.^2-6*x+1');
f3=inline('(((((x-6).*x+15).*x-20).*x+15).*x-6).*x+1');
k=0; n=100;
rng =[ .004 .003];
for delta=rng
  x=linspace(1-delta,1+delta,n);
  k=k+1; subplot(1,length(rng),k);
  y1=f1(x);
  y2=f2(x);
  y3=f3(x);
  plot(x,y1,'b*' , x, zeros(1,n),x,y2,'r-', x,y3,'g-')
  axis([1-delta 1+delta -max(abs(y2)),max(abs(y2))]);
end
```

5.3 *Fixed Point Example*

```
%  f(x)=x^3-4*x^2+3x+1 =0 has roots 2.8019,1.445,-0.24698 ;
% Fixed Point Solutions x in interval [2.5,4]
 g1=inline('(4*x.^2-3*x-1).^(1/3)');
 g2=inline('((1+3*x+x.^3)/4).^.5');
 g3=inline('4-1./x.^2-3./x');
 g1([2.5,4.0]);
 g2([2.5,4.0]);
 g3([2.5,4.0]);
   maxit=20;
p1=zeros(maxit,1);
p2=zeros(maxit,1);
p3=zeros(maxit,1);
p0=3.0;
p1(1,1)=p0;
p2(1,1)=p0;
p3(1,1)=p0;

  for k=1:maxit-1
    x=p1(k,1);
    p1(k+1,1)=g1(x);
    x=p2(k,1);
    p2(k+1,1)=g2(x);
    x=p3(k,1);
    p3(k+1,1)=g3(x);
  end
  format short g ;
k=[1:maxit]';
[k p1 p2 p3 ]
```

5.4 *Aitken's Convergence Acceleration*

```
%Aitken's Soln: f(x)=e^x-2*cos(x) =0 has roots
% Fixed Point Solutions x  [0,pi/4]
clear
g1=inline('log(2*cos(x))');
g2=inline('acos(exp(x)/2)');
g1([0,pi/4-eps]);
g2([0,pi/4-eps]);
maxit=31;
 p1=zeros(maxit,1);
 p2=zeros(maxit,1);
 p3=zeros(maxit,1);
 p0=0.25;
 p1(1,1)=p0;
 p2(1,1)=p0;

for k=1:maxit-1
   x=p1(k,1);
   p1(k+1,1)=g1(x);
   x=p2(k,1);
  p2(k+1,1)=g2(x);
end

  format short g;
  k=[1:maxit]';
  [k p1 p2 ]
  format long g;
[p1 [diff(p1);0] [diff(diff(p1));0;0] p1-[diff(p1);0].^2./[diff(diff(p1));0;0]]
```

5.5 *Taylor Polynomials for 1/x*

```
%TaylorPolyCompare
% Taylor polynomial for "1/x" about x0=1
% f(x)=1/x; f'(x)=-1/x^2;  f''(x)=+2/x^3;  f'''(x)=-6/x^4
%P1=f0+f0p(x-x0);
%P2=f0+f0p(x-x0)+f0pp(x-x0)^2/2;
%P3=f0+f0p(x-x0)+f0pp(x-x0)^2/2+f0ppp(x-x0)^3/6;

clear
x0=1; x=[.1:.1:5]; f=1./x;
f0=1./x0;
f0p=-(1./x0).^2;
f0pp=2*(1./x0).^3;
f0ppp=-6*(1./x0).^4;
P1=f0+f0p*(x-x0);
P2=P1+f0pp*(x-x0).^2/2;
P3=P2+f0ppp*(x-x0).^3/6;
figure(1);
axis equal;
plot(x,f,'k',x,P1,'r-',x,P2,'g:',x,P3,'b-',x0,f0,'r*');
axis([0 5 -10 10]);
title('Taylor Polynomials for  1/x about x_0=1');
xlabel('x');
ylabel('y=1/x');
grid on
```

5.6 Lagrange Polynomials for Sin(x) and Asin(x)

```
%Lagrange Interpolation
%Input Table for Sine Function
l1=[-pi/2:pi/5:pi/2];   f1=sin(l1);
lrng=[-2*pi:4*pi/100:2*pi];   frng=[-1:4/200:1];
syms f x L LI
nrng=length(lrng);   nrow =length(l1);
% Form Lagrange polynomial terms
fvec=x-l1;   fmtx=fvec;
for i=2:nrow
 fmtx=[fmtx;fvec];
end
onemtx= ones(nrow,nrow);              % nxn ones matrix
mask=onemtx-diag(ones(nrow,1));       % Zero out diagonal terms
tmp=fmtx.*mask+diag(ones(nrow,1));    % multiply by mask & put ones on diagonal
tmp2= prod(tmp,2);                    % Product forms Lagrange polynomial terms w/o coefficients
% Form denominators for polynomial coefficients
for i=1:nrow
tmp3(i)= subs(tmp2(i),x,l1(i));
end

L=sum((prod(tmp,2).*f1')./prod(tmp3',2))        % form product of fk*Lk and sum k=1,..,n
Lvpa=vpa(L,8)                                   % Lagrange polynomial  8 significant digits
% Form Inverse Lagrange Polynomials
fvec=x-f1; % Inverse Lagrange polynomial expands about "f1" values (rather than "l1")
fmtx=fvec;
for i=2:nrow
 fmtx=[fmtx;fvec];
end

tmp=fmtx.*mask+diag(ones(nrow,1));
tmp2= prod(tmp,2);
for i=1:nrow
tmp3(i)= subs(tmp2(i),x,f1(i));
end

LI=sum((prod(tmp,2).*l1')./prod(tmp3',2))
LIvpa=vpa(LI,8)                   % Inverse Lagrange polynomial  8 significant digits
% computing Lagrange Poly & Inverse & errors
delt=zeros(1,nrng);   deltI=zeros(1,nrng);   fL=zeros(1,nrng);   fLI=zeros(1,nrng);
delt1=zeros(1,nrow);   deltI1=zeros(1,nrow);   fL1=zeros(1,nrow);   fLI1=zeros(1,nrow);

for i=1:nrng
fL(i)=subs(L,x,lrng(i));
fLI(i)=subs(LI,x,frng(i));
delt(i) = fL(i)-sin(lrng(i));
deltI(i)= fLI(i)-asin(frng(i));
end

for i=1:nrow
fL1(i)=subs(L,x,l1(i));
fLI1(i)=subs(LI,x,f1(i));
delt1(i) = fL1(i)-sin(l1(i));
deltI1(i)= fLI1(i)-asin(f1(i));
end
figure(6)          % Composite plot Lagrange, sin, Error; Inverse Lagrange, asin, error
subplot(2,2,1)
```

```
plot(lrng,sin(lrng),'b-','LineWidth',2)
hold on
grid on
plot(lrng,fL,'r:','LineWidth',3)
plot(l1,sin(l1),'ko','MarkerSize',10,'MarkerFaceColor','k')
axis([-2*pi,2*pi,-1.5,1.5])
subplot(2,2,3)
semilogy(lrng,abs(delt),'r-','LineWidth',3)
hold on
axis([-2*pi,2*pi,10^-15,1])
grid on
subplot(2,2,2)
plot(frng,asin(frng),'b-','LineWidth',3)
hold on
grid on
plot(frng,fLI,'r:','LineWidth',4)
plot(f1,asin(f1),'ko','MarkerSize',10,'MarkerFaceColor','k')
axis([-1,1,-pi/2,pi/2])
subplot(2,2,4)
semilogy(frng,abs(deltI),'r-','LineWidth',3)
hold on
semilogy(f1,abs(deltI1),'ro','MarkerSize',10,'MarkerFaceColor','k')
grid on
axis([-1,1,10^-4,10^-1])
```

Outputs: *6 point Lagrangian* for *l1*=[-pi/2:pi/5:pi/2]

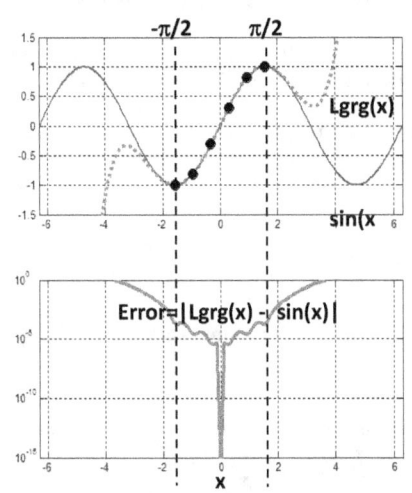
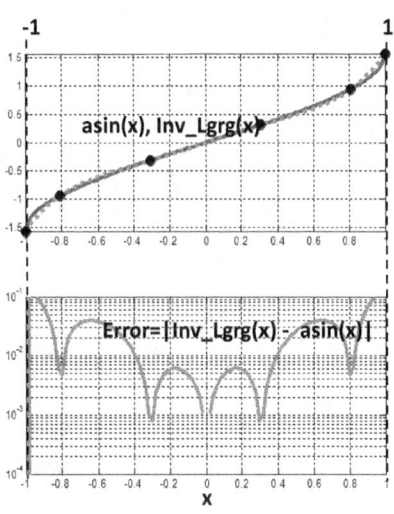

L(x) =.85098012e-1*(x+.94247781)*(x+.31415927)*(x-.31415927)*(x-.94247781)*(x-1.5707964)-
.34422869*(x+1.5707964)*(x+.31415927)*(x-.31415927)*(x-.94247781)*(x-
1.5707964)+.26296732*(x+1.5707964)*(x+.94247781)*(x-.31415927)*(x-.94247781)*(x-
1.5707964)+.26296732*(x+1.5707964)*(x+.94247781)*(x+.31415927)*(x-.94247781)*(x-1.5707964)-
.34422869*(x+1.5707964)*(x+.94247781)*(x+.31415927)*(x-.31415927)*(x-1.5707964)+.85098012e-
1*(x+1.5707964)*(x+.94247781)*(x+.31415927)*(x-.31415927)*(x-.94247781)

LI(x) =2.5132742*(x+.80901699)*(x+.30901699)*(x-.30901699)*(x-.80901699)*(x-1.)-
3.0159290*(x+1.)*(x+.30901699)*(x-.30901699)*(x-.80901699)*(x-1.)+1.0053097*(x+1.)*(x+.80901699)*(x-
.30901699)*(x-.80901699)*(x-1.)+1.0053097*(x+1.)*(x+.80901699)*(x+.30901699)*(x-.80901699)*(x-1.)-
3.0159290*(x+1.)*(x+.80901699)*(x+.30901699)*(x-.30901699)*(x-
1.)+2.5132742*(x+1.)*(x+.80901699)*(x+.30901699)*(x-.30901699)*(x-.80901699)

MatLab Scripts

Outputs: *11 point Lagrangian* for l1=[-pi/2:pi/10:pi/2]

Lagrange Polynomial for sin(x)

L(x) = -40.212387*(x+.95105652)*(x+.80901699)*(x+.58778525)*(x+.30901699)*x*(x-.30901699)*(x-.58778525)*(x-.80901699)*(x-.95105652)*(x-1.)+64.339818*(x+1.)*(x+.80901699)*(x+.58778525)*(x+.30901699)*x*(x-.30901699)*(x-.58778525)*(x-.80901699)*(x-.95105652)*(x-1.)-48.254864*(x+1.)*(x+.95105652)*(x+.58778525)*(x+.30901699)*x*(x-.30901699)*(x-.58778525)*(x-.80901699)*(x-.95105652)*(x-1.)+32.169909*(x+1.)*(x+.95105652)*(x+.80901699)*(x+.30901699)*x*(x-.30901699)*(x-.58778525)*(x-.80901699)*(x-.95105652)*(x-1.)-16.084955*(x+1.)*(x+.95105652)*(x+.80901699)*(x+.58778525)*x*(x-.30901699)*(x-.58778525)*(x-.80901699)*(x-.95105652)*(x-1.)+16.084955*(x+1.)*(x+.95105652)*(x+.80901699)*(x+.58778525)*(x+.30901699)*x*(x-.58778525)*(x-.80901699)*(x-.95105652)*(x-1.)-32.169909*(x+1.)*(x+.95105652)*(x+.80901699)*(x+.58778525)*(x+.30901699)*x*(x-.30901699)*(x-.80901699)*(x-.95105652)*(x-1.)+48.254864*(x+1.)*(x+.95105652)*(x+.80901699)*(x+.58778525)*(x+.30901699)*x*(x-.30901699)*(x-.58778525)*(x-.95105652)*(x-1.)-64.339818*(x+1.)*(x+.95105652)*(x+.80901699)*(x+.58778525)*(x+.30901699)*x*(x-.30901699)*(x-.58778525)*(x-.80901699)*(x-1.)+40.212387*(x+1.)*(x+.95105652)*(x+.80901699)*(x+.58778525)*(x+.30901699)*x*(x-.30901699)*(x-.58778525)*(x-.80901699)*(x-.95105652)

Inverse Lagrange Polynomial for asin(x)

LI(x) = -40.212387*(x+.95105652)*(x+.80901699)*(x+.58778525)*(x+.30901699)*x*(x-.30901699)*(x-.58778525)*(x-.80901699)*(x-.95105652)*(x-1.)+64.339818*(x+1.)*(x+.80901699)*(x+.58778525)*(x+.30901699)*x*(x-.30901699)*(x-.58778525)*(x-.80901699)*(x-.95105652)*(x-1.)-48.254864*(x+1.)*(x+.95105652)*(x+.58778525)*(x+.30901699)*x*(x-.30901699)*(x-.58778525)*(x-.80901699)*(x-.95105652)*(x-1.)+32.169909*(x+1.)*(x+.95105652)*(x+.80901699)*(x+.30901699)*x*(x-.30901699)*(x-.58778525)*(x-.80901699)*(x-.95105652)*(x-1.)-16.084955*(x+1.)*(x+.95105652)*(x+.80901699)*(x+.58778525)*x*(x-.30901699)*(x-.58778525)*(x-.80901699)*(x-.95105652)*(x-1.)+16.084955*(x+1.)*(x+.95105652)*(x+.80901699)*(x+.58778525)*(x+.30901699)*x*(x-.58778525)*(x-.80901699)*(x-.95105652)*(x-1.)-32.169909*(x+1.)*(x+.95105652)*(x+.80901699)*(x+.58778525)*(x+.30901699)*x*(x-.30901699)*(x-.80901699)*(x-.95105652)*(x-1.)+48.254864*(x+1.)*(x+.95105652)*(x+.80901699)*(x+.58778525)*(x+.30901699)*x*(x-.30901699)*(x-.58778525)*(x-.95105652)*(x-1.)-64.339818*(x+1.)*(x+.95105652)*(x+.80901699)*(x+.58778525)*(x+.30901699)*x*(x-.30901699)*(x-.58778525)*(x-.80901699)*(x-1.)+40.212387*(x+1.)*(x+.95105652)*(x+.80901699)*(x+.58778525)*(x+.30901699)*x*(x-.30901699)*(x-.58778525)*(x-.80901699)*(x-.95105652)

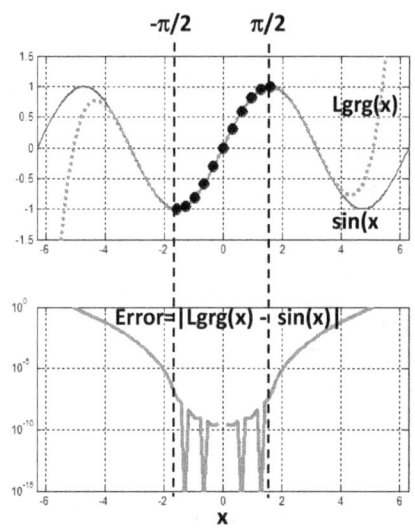

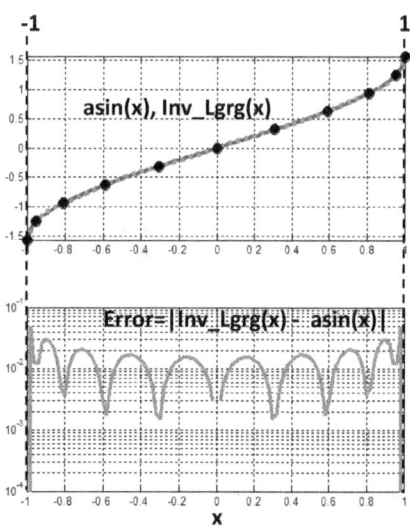

5.7 *Richardson Extrapolation*

```
% Richardson Extrapolation for Numerical Derivative
clear
d=8;
digits(d); % set computer wordsize "digits" d=5
f1=inline('exp(x)'); %Define exponential function
k=8:12; % vector to "loop over" for truncated table
hk=2.^-k;% halve the "h-value": 2 raised to the "vector" -k
hk=vpa(hk); %symbolic round off to d digits
Dk=zeros(length(k),1);
Dk=vpa((vpa(f1(1+hk))-vpa(f1(1)))./hk);
delta=abs(vpa(Dk)-vpa(f1(1)));

jump=1; % "jump" term-by-term jump in order of h (stepsize structure)
      %e.g. M=N1[h]+a1*h^1+ a2*h^2+ a3*h^3+... Linear jump=1
      % M=N1[h]+a1*h^2+a2*h^4+ a3*h^6+... Quadratic jump=2
Dk=double(Dk);
delta=double(delta);
hk=double(hk);
format short g
{ 'hk' '1st Deriv'}

[hk',Dk']
lcol=length(Dk');
extrap_tbl=zeros(lcol);
extrap_tbl(:,1)=Dk';
extrap0=Dk';
for j=2:lcol
denom=(2^jump)^(j-1)-1;
addtemp=diff(extrap_tbl(j-1:lcol,j-1))/denom;
extrap_tbl(j:lcol,j)=extrap_tbl(j:lcol,j-1)+addtemp;
end
format long g
[k' hk' extrap_tbl]

plotflag=0;
if plotflag==1
figure(1)
hold on
plot(-log2(hk),Dk,'*-')
plot(-log2(hk), repmat(exp(1),1,length(hk)),'r:','linewidth',2)
%plot(-log2(hk), log10(delta),'g:')
%hold off
end
```

5.8 *Bezier Curves*

```
%Bezier Curves
%Bezier Curves Comparison
clear all
xL=3;yL=4;     %Fixed Left Guide Point
t=linspace(0,1); %Parametric t-values for plot
x=inline('xL*t+(1-xL)*t.^2+(xL+(xR-1)-2)*(t-1).*t.^2','t','xL','xR');   %x(t)
y=inline('yL*t-yL*t.^2+(yL+yR)*(t-1).*t.^2','t','yL','yR');          %y(t)

figure(1)
hold on
grid on
xR=-3;yR=-2;   %Right Guide Point #1
plot(xL,yL,'r*',xR,yR,'g*')
line([xL,0],[yL,0],'color',[.8,.8,.8],'linewidth',2)
line([xR,1],[yR,0],'color',[.8,.8,.8],'linewidth',2)
plot(x(t,xL,xR),y(t,yL,yR),'b--','linewidth',4)

xR=5;yR=-2;   %Right Guide Point #2
plot(xL,yL,'r*',xR,yR,'g*')
line([xL,0],[yL,0],'color',[.8,.8,.8],'linewidth',2)
line([xR,1],[yR,0],'color',[.8,.8,.8],'linewidth',2)
plot(x(t,xL,xR),y(t,yL,yR),'r','linewidth',4)

xR=-3;yR=2;   %Right Guide Point #3
plot(xL,yL,'r*',xR,yR,'g*')
line([xL,0],[yL,0],'color',[.8,.8,.8],'linewidth',2)
line([xR,1],[yR,0],'color',[.8,.8,.8],'linewidth',2)
plot(x(t,xL,xR),y(t,yL,yR),'g:','linewidth',4)

xR=-3;yR=5;   %Right Guide Point #4
plot(xL,yL,'r*',xR,yR,'g*')
line([xL,0],[yL,0],'color',[.8,.8,.8],'linewidth',2)
line([xR,1],[yR,0],'color',[.8,.8,.8],'linewidth',2)
plot(x(t,xL,xR),y(t,yL,yR),'g','linewidth',4)
```

5.9 Bezier Curves - Interactive

```
%Interactive Bezier Curves
clear
axis([-5 5 -5 5])
figure(1)
hold on
grid on
t=linspace(0,1);

but = 1;   % set button flag equal to 1 for interactive mouse clicks
while but == 1
    [xL,yL,but] = ginput(1); % click mouse at left guide point
    plot(xL,yL,'r*')         % plot left guide point green
    [xR,yR,but] = ginput(1); % click mouse at right guide point
    plot(xR,yR,'g*')         % plot right guide point red
    xR1=xR-1;
x=inline('xL*t+(1-xL)*t.^2+(xL+xR1-2)*(t-1).*t.^2','t','xL','xR1'); %parametric x(t)
y=inline('yL*t-yL*t.^2+(yL+yR)*(t-1).*t.^2','t','yL','yR');         %parametric y(t)
plot(x(t,xL,xR1),y(t,yL,yR))                        % plot bezier curve
line([xL,0],[yL,0],'color',[.8,.8,.8])        %draw left guide line
line([xR1+1,1],[yR,0],'color',[.8,.8,.8])       %draw right guide line
end
hold off
```

5.10 *Root Finding Methods Comparison*

```
RootFindingMethodsCompare
clear all
f=inline('x^3-2*x^2-5');
fp=inline('3*x^2-4*x');
fpp=inline('6*x-4');
g1=inline('(2*x^2+5)^(1/3)');
g2=inline('2+5/x^2');
%g3=inline('((x^3-5)/2)^(1/2)');
a=2.0; %a=-12.0;%Muller
b=2.5;%Muller
c=3.0;%Muller
xnr=a;%Newton-Raphson
xfp=a;% & Fixed Point
x0=a;% Secant
x1=b;%Secant
x2=c;%Muller
imax=20;
out=zeros(imax+1,10);

%Secant Method
out(1,1)=x0;
out(2,1)=x1;
%out(1:2,2)=0;
for i=3:imax+2
corr=-f(x0)*(x1-x0)/(f(x1)-f(x0));
xnew=x0+corr;
x0=x1;
x1=xnew;
out(i,1)=x1;
end

%Newton-Raphson Method
xnr=a;
out(1,2)=xnr;
for i=2:imax+2
corr=-f(xnr)/fp(xnr);
xnew=xnr+corr;
xnr=xnew;
out(i,2)=xnr;
end

%Bisection
an=a;
bn=c;
out(1,3)=an; %a
out(1,4)=bn; %b
```

```
out(1,5)=0;

for i=2:imax+2
pn=(an+bn)/2;
out(i,5)=pn;

fpn=f(pn);
fan=f(an);
fbn=f(bn);
  if fpn*fbn<0
  an=pn;
  out(i,3)=an;
  out(i,4)=bn;

  else   %(if fpn*fan<0)
  bn=pn;
  out(i,3)=an;
  out(i,4)=pn;
  end
end

%Fixed Point Methods
xfp1=a;xfp2=a;xfp3=a;
out(1,6)=xfp1;out(1,7)=xfp2;
for i=2:imax+2
xfp1=g1(xfp1);
xfp2=g2(xfp2);
%xfp3=g3(xfp3);

out(i,6)=xfp1;out(i,7)=xfp2;
end

% Tschebyshev Method
out(1,8)=a;
xch=a;
for i=2:imax+2
corr=-(f(xch)/fp(xch))*(1+f(xch)*fpp(xch)/(2*fp(xch)^2));
xch=xch+corr;
out(i,8)=xch;
end

% Cauchy Method
out(1,9)=a;
xca=a;
for i=2:imax+2
corr=-(2*f(xca)/fp(xca))/(1+sqrt(1-2*f(xca)*fpp(xca)/fp(xca)^2));
xca=xca+corr;
out(i,9)=xca;
end
```

MatLab Scripts

```
%Muller Method
out(1,10)=a;
out(2,10)=b;
out(3,10)=c;
x0=a; x1=b;x2=c;
for i=4:imax+2
a1=(-(x0-x2)*(f(x1)-f(x2))+(x1-x2)*(f(x0)-f(x2)))/((x0-x2)*(x1-x2)*(x0-x1));
b1=((x0-x2)^2*(f(x1)-f(x2))-(x1-x2)^2*(f(x0)-f(x2)))/((x0-x2)*(x1-x2)*(x0-x1));
c1=f(x2);
xnew=x2-2*c1/(b1+sign(b1)*sqrt(b1^2-4*a1*c1));
x0=x1;
x1=x2;
x2=xnew;
out(i,10)=xnew;
end
format short;
'                         BISECTION      Fixed Point '
'  Secant NwtnRaphs      an       bn      pn       g1      g2      Cheb      Cauchy   Muller'
 out

out2(:,1)=out(:,1);out2(:,2)=out(:,2);out2(:,3)=out(:,8);out2(:,4)=out(:,9);out2(:,5)=out(:,10);
format long;
'                         BISECTION      Fixed Point '
'  Secant           NwtnRaphs         Cheb          Cauchy          Muller'
 out2
figure(1)
ezplot(f);hold on
ezplot(g1)
ezplot(g2)
ezplot(g3)
grid on
figure(2)
x=linspace(0,20,1000);
for k=1:length(x)
   ff(k,1)=f(x(k));
   gg1(k,1)=g1(x(k));
   gg2(k,1)=g2(x(k));
plot(x,ff,'r-')
hold on
plot(x,x,'b-')
plot(x,gg1,'b:')
plot(x,gg2,'g--')
plot(x,gg3,'r--')
title('red ff=x^3-2*x^2-5; blue y=x; blue dot gg1; green dash gg2; red dash gg3')
axis([0,20,-10,10])
grid on
end
```

5.11 *Simpson 1^d Adaptive Quadrature*

```
%AdaptiveQuadrature wkg_10z1
np=1;   % one Simp panel
out=zeros(nmax+1,7);
%outjvec={};
flags=[1, 1];%{flagL flagR} L, R continue flags
Itmp=0;   % initialize Integral value at zero Itmp
stopflg=0;
jvec=[1,2];  %active panels
Itmpj=0;       % sum of extrapolations for given split level
xx=zeros(1,9);
yy=zeros(1,9);
xx=[a,b];yy=[f(a),f(b)];
            %ezplot('(100./x.^2).*sin(10./x)', [a,b])
             ezplot(eq, [a,b]) % Plots chosen (ex==) equation "eq"
             grid on
             hold on
             plot(xx,yy,'k*')


%@@@@@@@@@@@@@@@@@@@@@@@@@@@@@@@
   if abs((I_s1(a,b,del0,f)-I_s2(a,b,del0/2,f))/15) <= e_tol  % initial test
     Itmp=extrap(a,b,h,f);
     %sum(flags,2)=0;
      stopflg=1;
   end
out(1,1)=Itmp; out(1,2)=Itmpj;out(1,3)=e_tol;out(1,4)=np;out(1,5)=h;out(1,6)=size(flags,2);
out(1,7)=sum(flags,2);%output integral
   if stopflg==1
                   '   Itmp    Itmpj correction   e_tol     np panels before split     new h     double
panels    active panels '
     out
     xx=[a,a+h,b];yy=[f(a),f(a+h),f(b)];
     plot(xx,yy,'ro')
   end
%@@@@@@@@@@@@@@@@@@@@@@@@@@@@@@@@@@@@@@@@@@@@@@@@@@@@@
if stopflg == 0 %Stops here if the 1-panel Simpson satisfies target e_tol
flags1=[] ;  %initialize flags1 vector
for ns=1:nmax      % begin split loop h -> h/2
   if sum(flags,2) > 0 % test active panel
      h=h/2 ;        % half previous h for L & R   panels
      e_tol=e_tol/2 ; % half e_tol for L & R  panels
      np=2^ns;% total # panels np
      deltaj= h;   %(b-a)/2^ns;
      jvec=find(flags.*[1:2^ns]); %active panels NOT EXTRAPOLATED
      flags1=[]; % "0/1" mask accumulator
      Itmpj=0;
      for j = 1:2^ns
        aj=a+(j-1)*2*h;
        bj=a+j*2*h;
```

```matlab
        flagsj=[];% "0/1" reset j loop accumulator flagsj
        xxj=[];yyj=[];
      if j==jvec(find(jvec==j)) % j==find(jvec==j); %gives zero if j not in jvec; no extrap.
            IS= I_s1(aj,bj,deltaj,f) ; %one panel for current [aj, bj] panel
            IS2=I_s2(aj,bj,deltaj/2,f);   % two panel Simpson with h/2
             test=(IS-IS2)/15 ;
          if abs(test) <= e_tol;            %  test of one panel
            IE=extrap(aj, bj, deltaj, f) ;    % two panel Extrap
            Itmpj=Itmpj+IE;
            xxj=[aj,aj+h,aj+2*h] ;
            yyj=[f(aj),f(aj+h),f(aj+2*h)];
            plot(xxj,yyj,'ro')
            xxab=[aj,bj];
            yy0=[0,0];
            plot(xxab,yy0,'r.')
            %f(a)+4*f((a+del))+2*f(a+2*del)+4*f(a+3*del)+f(b)
            flags(j)=0 ; %no longer active
            flagsj=[repmat(flags(j),1,2)]; % double  0 0  for next split
          else %  if not meeting e_tol flag =1
            flags(j)=1 ; %remains active
            flagsj=[repmat(flags(j),1,2)] ; % double  1 1   for next split
          end        % end if abs(test) ;
      else           %if j not a member of jvec
            flags(j)=0 ; %zeros remain inactive & double for next split ns value
            flagsj=[repmat(flags(j),1,2)] ;% double  0 0
      end            % end j member of jvec   active
            flags1=[flags1 flagsj];      % adding flagsj to 1/0  mask flags1
    end   % end for j loop
      Itmp=Itmp+Itmpj;    % increment Itmp at end of j-loop
      flags=flags1 ;     % final mask flags for next split ns to compute new jvec
      %out(ns+1,1)=Itmp; out(ns+1,2)=e_tol;out(ns+1,3)=np;out(ns+1,4)=h;out(ns+1,5)=size(flags,2);
out(ns+1,6)=sum(flags,2);out(ns+1,7)=Itmpj;
        out(ns+1,1)=Itmp;
out(ns+1,2)=Itmpj;out(ns+1,3)=e_tol;out(ns+1,4)=np;out(ns+1,5)=h;out(ns+1,6)=size(flags,2);
out(ns+1,7)=sum(flags,2);%output integral
            %jvec %diagnostic
    end        % end of if sum(flags,2) > 0 % test active panel
  end           % end for ns split LR loop
          if sum(flags,2)==0
              hold off
          %' Itmp  Itmpj   e_tol         np             h          size(flags,2)      sum(flags,2)'
          ' Itmp   Itmpj   e_tol #panels_old  h_ new   #panels_new   #panels_ active '
              out
              format long g
        Itmpj
              Itmp
              end
  end        %  end stopflg loop
```

ex==9   eq='sin(50.*x)./(pi*x)';a=-0.1;b=0.1; e_tol = 1e-005

Itmp	Itmpj	e_tol	np	h	size(flags,2)	sum(flags,2)
Itmp	tmpj	e_tol	#panels_old	h_new	#panels_new	#panels_active
0	0	1e-005	1	0.1	2	2
0	0	5e-006	2	0.05	4	4
0	0	2.5e-006	4	0.025	8	8
0.40239	0.40239	1.25e-006	8	0.0125	16	12
0.59737	0.19498	6.25e-007	16	0.00625	32	4
0.78885	0.19149	3.125e-007	32	0.003125	64	4
0.88738	0.09853	1.5625e-007	64	0.0015625	128	4
0.937	0.049618	7.8125e-008	128	0.00078125	256	4
0.96185	0.024853	3.9063e-008	256	0.00039063	512	4
0.97429	0.012432	1.9531e-008	512	0.00019531	1024	4
0.9805	0.0062168	9.7656e-009	1024	9.7656e-005	2048	4

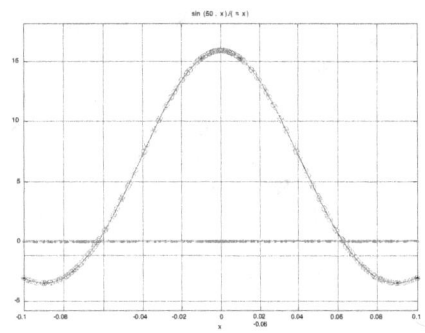

ex==9   eq='sin(50.*x)./(pi*x)';a=-1.1;b=1.1; e_tol = 1e

Itmp	Itmpj	e_tol	#panels_old	h_new	#panels_new	#panels_active
0	0	1e-005	1	1.1	2	2
0	0	5e-006	2	0.55	4	4
0	0	2.5e-006	4	0.275	8	8
0	0	1.25e-006	8	0.1375	16	16
0	0	6.25e-007	16	0.06875	32	32
0	0	3.125e-007	32	0.034375	64	64
-0.0035068	-0.0035068	1.5625e-007	64	0.017188	128	124
-0.086733	-0.083226	7.8125e-008	128	0.0085938	256	120
0.7292	0.81593	3.9063e-008	256	0.0042969	512	4
0.86353	0.13433	1.9531e-008	512	0.0021484	1024	4
0.93161	0.06808	9.7656e-009	1024	0.0010742	2048	4

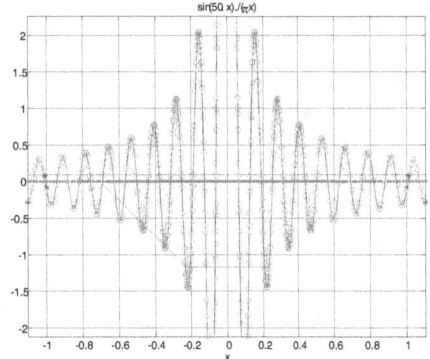

ex==9   eq='sin(50.*x)./(pi*x)';a=-50;b=50; e_tol = .00

Itmp	Itmpj	e_tol	#panels_old	h_new	#panels_new	#panels_active
0	0	0.001	1	50	2	2
0	0	0.0005	2	25	4	4
0	0	0.00025	4	12.5	8	8
0.094533	0.094533	0.000125	8	6.25	16	12
0.094533	0	6.25e-005	16	3.125	32	24
0.092865	-0.0016681	3.125e-005	32	1.5625	64	44
0.072033	-0.020832	1.5625e-005	64	0.78125	128	56
0.072033	0	7.8125e-006	128	0.39063	256	112
0.072102	6.8653e-005	3.9063e-006	256	0.19531	512	216
0.072102	0	1.9531e-006	512	0.097656	1024	432
0.073442	0.0013405	9.7656e-007	1024	0.048828	2048	832
0.072797	-0.00064537	4.8828e-007	2048	0.024414	4096	976
0.40008	0.32728	2.4414e-007	4096	0.012207	8192	148
0.69235	0.29227	1.2207e-007	8192	0.0061035	16384	4
0.87968	0.18733	6.1035e-008	16384	0.0030518	32768	4
0.97594	0.096264	3.0518e-008	32768	0.0015259	65536	4
1.0244	0.04846	1.5259e-008	65536	0.00076294	1.31e+005	4

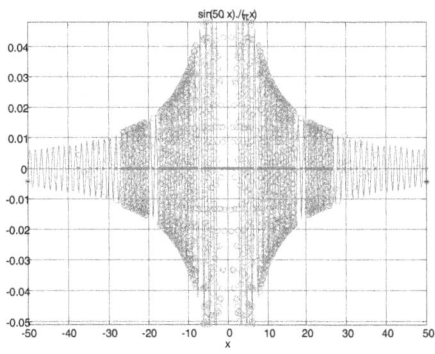

ex==1 "BF" eq='(100./x.^2).*sin(10./x)'; a=1;b=3; **e_tol = .001**

Itmp sum(flags,2)	Itmpj	e_tol	np	h	size(flags,2)	
Itmp	Itmpj	e_tol	#panels$_{old}$	h$_{new}$	#panels$_{new}$	#panels$_{active}$
0	0	0.001	1	1	2	2
0	0	0.0005	2	0.5	4	4
-3.2807	-3.2807	0.00025	4	0.25	8	6
-18.243	-14.962	0.000125	8	0.125	16	6
-15.308	2.9345	6.25e-005	16	0.0625	32	6
-1.4268	13.881	3.125e-005	32	0.03125	64	0

Itmp =   -1.426827727713   Itmpj =   13.8812724093772

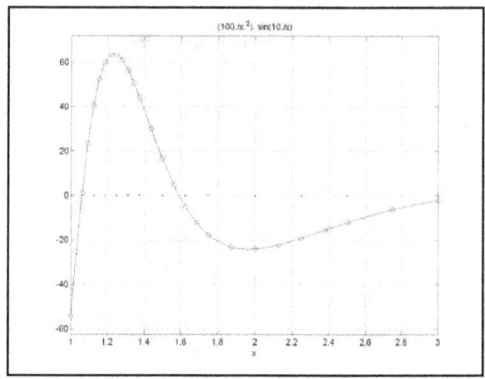

ex==1"BF" eq='(100./x.^2).*sin(10./x)'; a=1;b=3; **e_tol = .0001**

Itmp sum(flags,2)	Itmpj	e_tol	np	h	size(flags,2)	
Itmp	Itmpj	e_tol	#panels$_{old}$	h$_{new}$	#panels$_{new}$	#panels$_{active}$
0	0	0.0001	1	1	2	2
0	0	5e-005	2	0.5	4	4
0	0	2.5e-005	4	0.25	8	8
-3.2803	-3.2803	1.25e-005	8	0.125	16	12
-19.733	-16.453	6.25e-006	16	0.0625	32	10
-4.4451	15.288	3.125e-006	32	0.03125	64	6
-1.426	3.0191	1.5625e-006	64	0.015625	128	0

Itmp =   -1.42604625428452   Itmpj =   3.01906558799358

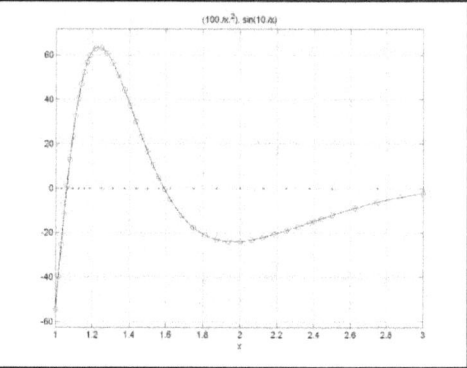

ex 7) '2*x.*cos(2*x)-(x-2).^2' ;a=0;b=5; e_tol = 1e-005

Itmp	Itmpj	e_tol	# panel	h_{new}	#panels$_{new}$	#panels$_{active}$
0	0	1e-005	1	2.5	2	2
0	0	5e-006	2	1.25	4	4
0	0	2.5e-006	4	0.625	8	8
0	0	1.25e-006	8	0.3125	16	16
0.79355	0.79355	6.25e-007	16	0.15625	32	30
-9.9031	-10.697	3.125e-007	32	0.078125	64	4
-15.306	-5.4032	1.5625e-007	64	0.039063	128	0

Itmp = -15.3063093510881 ; e_tol = 1e-005   64 panels

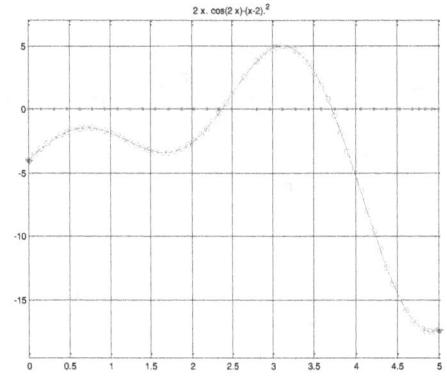

ex 7) '2*x.*cos(2*x)-(x-2).^2' ;a=0;b=5; e_tol = 1e-007

Itmp	Itmpj	e_tol	# panel	h_{new}	#panels$_{new}$	#panels$_{active}$
0	0	1e-007	1	2.5	2	2
0	0	5e-008	2	1.25	4	4
0	0	2.5e-008	4	0.625	8	8
0	0	1.25e-008	8	0.3125	16	16
0	0	6.25e-009	16	0.15625	32	32
0	0	3.125e-009	32	0.078125	64	64
-1.7587	-1.7587	1.5625e-009	64	0.039063	128	10
-15.306	-13.548	7.8125e-010	128	0.019531	256	0

Itmp = -15.3063079957494; e_tol = 1e-007   128 panels

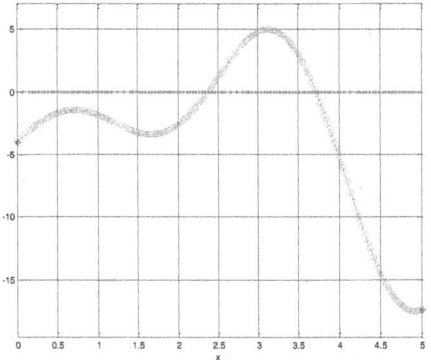

ex 7) '2*x.*cos(2*x)-(x-2).^2' ;a=0;b=5; e_tol = 1e-010

Itmp	Itmpj	e_tol	# panel	h_{new}	#panels$_{new}$	#panels$_{active}$
0	0	1e-010	1	2.5	2	2
0	0	5e-011	2	1.25	4	4
0	0	2.5e-011	4	0.625	8	8
0	0	1.25e-011	8	0.3125	16	16
0	0	6.25e-012	16	0.15625	32	32
0	0	3.125e-012	32	0.078125	64	64
0	0	1.5625e-012	64	0.039063	128	128
0.10684	0.10684	7.8125e-013	128	0.019531	256	254
-0.58296	-0.6898	3.9063e-013	256	0.0097656	512	492
-8.6099	-8.0269	1.9531e-013	512	0.0048828	1024	230
-15.306	-6.6964	9.7656e-014	1024	0.0024414	2048	0

Itmp = -15.3063079856533;e_tol = 1e-010 1024 panels

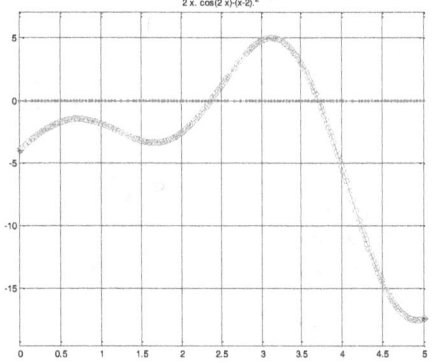

5.12 Gauss 1^d Multi-Panel Quadrature

MultiPanel_GaussQuadr.m

```matlab
%MultiPanel_GaussQuadr.m
% From AdaptQuadr_Gauss_Book_wkg7_OK.m
clear all
format short g
' *** MultiPanel_GaussQuadr *** '
%@@@@@@@@@@ Examples @@@@@@@@@@@@@
%0) f=inline('x./sqrt(4*x.*(1-x))'); a=0.0 ; b=1.0;
%1)f=inline('(100./x.^2).*sin(10./x)');a=1;b=3;
%2)f=inline('exp(2.*x).*sin(3*x)');a=1;b=3;
%3)f=inline('exp(3.*x).*sin(2*x)');a=1;b=3;
%4) f=inline('x.^5');a=1;b=3;
%6)f=inline('sin(x)');a=0; b=pi/2;
%7) f=inline('2*x.*cos(2*x)-(x-2).^2');a=0;b=5;
%8) f=inline('exp(3.*x).*sin(2*x.^2)');a=2;b=4;
%9) f=inline('sin(50.*x)./(pi*x)');a=-100;b=100; %Approx delta function)
%10 f=inline('1./sqrt(1-3*sin(x).^2)'); a=0.0 ; b=2.0;
%@@@@@@@@@@@@@@@@@@@@@@@@@@@@@@@
nmax=12; % max splits of panels
n1=10; %n1-point Gaussian
e_tol0=1e-4; % Target Accuracy
I_ng=0;
pltflg=zeros(1,10);
%flag# 1 2 3 4 5 6 7 8 9 10
pltflg=[1,1,1,1,0,0,1,1,0,0];
ex=1 ; % **** Choose examples 0 - 10 ****

if ex==0
    eq='x./sqrt(4*x.*(1-x))'; a=0.0; b=1.0;I_exact =pi/4;% Exact=pi/4=0.785398163397448;HP Prime 0.785398106639
elseif ex==1
    eq='(100./x.^2).*sin(10./x)'; a=1;b=3; I_exact =-1.42602475635;% HP Prime -1.42602475635
 elseif ex==2
    eq='exp(2.*x).*sin(3*x)';a=1;b=3; %BF4.6-3a HP Prime
 elseif ex==3
    %eq='exp(3.*x).*sin(2*x)';a=1;b=3; I_exact = -1724.96698301 % 3a HP Prime [1,3]: -1724.96698301
    eq='exp(3.*x).*sin(2*x)';a=2;b=5; I_exact= 11617.2287359 ; % 3b HP Prime [2,5]: 11617.2287359 ; [1,5]:
11581.8396063
 elseif ex==4
    eq= 'x.^9';a=1;b=3; I_exact=5904.8 ;        % HP Prime 5904.8 exact soln
    %eq= 'x.^3';a=1;b=3;            % HP Prime 5904.8 exact soln
 elseif ex==5
    eq='sin(x)./x';a=-.1;b=.1001; %BF4.6-6a,b HP Prime
 elseif ex==6
    eq='x.*cos(x.^2)';a=0; b=pi;%BF4.6-5a HP Prime
 elseif ex==7
    eq='2*x.*cos(2*x)-(x-2).^2';a=0;b=5;%%BF4.6-3c HP Prime
 elseif ex==8
  %eq='exp(3.*x).*sin(2*x.^2)';a=2;b=4; I_exact= -7339.32745549; %8a HP Prime   -7339.32745549
   eq='exp(3.*x).*sin(2*x.^2)';a=2;b=5; I_exact= -160808.749439; %8b HP Prime [2,5] -160808.749439
;32785386.9341
 elseif ex==9
    eq='sin(50.*x)./(pi*x)';a=-1;b=1.00001;I_exact=1; %Approx delta function
    %eq='sin(50.*x)./(pi*x)';a=-5;b=1.00001; %Approx delta function
 elseif ex==10
    eq=' 1./sqrt(1-3*sin(x).^2)'; a=0.0 ; b=2.0; %HP Prime 1.00107718634 - 1.49027809758 *i
end
```

MatLab Scripts

```
f=inline(eq,'x');
e_tol=e_tol0;
m1=n1-1;   % for Gauss Roots and Coeff look-up
coef(1,1).rc=[ -1/sqrt(3)  1/sqrt(3); 1 1] ; %2 pt Gauss Quadr
coef(1,2).rc=[ -sqrt(3/5) sqrt(3/5) 0 ; 5/9  5/9 8/9]; %3 pt Gauss Quadr
coef(1,3).rc=[-sqrt((15+2*sqrt(30))/35) sqrt((15+2*sqrt(30))/35) -sqrt((15-2*sqrt(30))/35) sqrt((15-2*sqrt(30))/35) ;
    (18-sqrt(30))/36  (18-sqrt(30))/36 (18+sqrt(30))/36 (18+sqrt(30))/36];% 4 pt Gauss Quadr
coef(1,4).rc=[-0.5384693101056381 0.5384693101056381 -0.9061798459386640  0.9061798459386640 0.0 ;
    0.4786286704993665 0.4786286704993665 0.2369268850561891 0.2369268850561891 .568888888888889];% 5 pt
Gauss Quadr
coef(1,5).rc=[-0.6612093864662645 0.6612093864662645 -0.2386191860831969 0.2386191860831969 -
0.9324695142031521 0.9324695142031521;
    0.3607615730481386 0.3607615730481386 0.4679139345726910 0.4679139345726910 0.1713244923791704
0.1713244923791704 ];%6 pt Quadr
coef(1,6).rc=[-0.4058451513773972 0.4058451513773972 -0.74155311855993945 0.74155311855993945 -
0.9491079123427585 0.9491079123427585 0.0;
0.3818300505051189  0.3818300505051189 0.2797053914892766 0.2797053914892766 0.1294849661688697
0.1294849661688697 0.4179591836734694 ]; %7 pt
coef(1,7).rc=[-0.18343464249565 0.18343464249565 -0.525532409916329 0.525532409916329 -0.796666477413627
0.796666477413627 -0.960289856497536 0.960289856497536;
0.362683783378362 0.362683783378362 0.313706645877887 0.313706645877887 0.222381034453375
0.222381034453375 0.101228536290376 0.101228536290376];%8pt
coef(1,8).rc=[0 -0.836031107326636 0.836031107326636 -0.968160239507626 0.968160239507626 -
0.324253423403809 0.324253423403809 -0.61337143270059 0.61337143270059;
0.33023935500126 0.180648160694857 0.180648160694857 0.0812743883615744 0.0812743883615744
0.312347077040003 0.312347077040003 0.260610696402935 0.260610696402935];%9pt
coef(1,9).rc=[-0.148874339 0.148874339 -0.433395394 0.433395394 -0.679409568 0.679409568 -0.865063367
0.865063367 -0.973906529 0.973906529
0.295524225     0.295524225     0.269266719     0.269266719     0.219086363     0.219086363     0.149451349
      0.149451349     0.066671344     0.066671344 ]; %10 pt


% a,b to -1,1 Transf
h1=(b-a)/2; h2=(b+a)/2 ;          % h1 = initial xcale factor from f(x) to g(t) ;  h2 = "x-coordinate origin shift"
u=coef(1,m1).rc(1,:) ;          %  u root values 'rn1i'
x=h1*u+h2 ;               % x [a,b]  to t {-1 ,1] xform
cn1i=coef(1,m1).rc(2,:) ;         % Gauss Quadra coeficients
fi=f(x) ;
gi=h1.*fi ;
I_g=sum(cn1i.*gi) ;
format long g
{ 'ex', 'eq','a','b','e_tol','nmax', 'nptGauss','I_exact'; ex,eq,a, b,e_tol0,nmax, n1,I_exact}
outg = [ 0 1 I_g I_g];   %outg = [ ns np I_g I_exact];
%  *******e*********
convflg = 0;
ns=0;
nconv=0;

while ns <= nmax & convflg ==0
    ns=ns+1;
    term1=0; term2=0;term3=0;
    alpha =[1:2:2^ns-1];
    pts=[-1:1/2^(ns-1):1];
    npts=length(pts);
    del_ns=(b-a)/2^ns;
    hh1=(pts(npts)-pts(npts-1))/2 ;
    hh2=(pts(npts)+pts(npts-1))/2 ;
    np=2^ns;
    del_h =(b-a)/(2*np);
```

```matlab
    for kk=1:(npts-1)/2
      gxx1=h1*hh1*( f(h1*hh1*(u-alpha(kk))+h2) + f(h1*hh1*(u+alpha(kk))+h2) );%The "h2" L & R shift by the original
integral "c"s and "u"s
      term1= term1 + sum(cn1i.*gxx1);
    end

    for kk=1:np
      gxx2=del_h*f( del_h*(u + (2*kk-1)) + a );          % h1*f((h1/np)*(u+(2*kk-1)+a)+ h2)
      term2= term2 + sum(cn1i.*gxx2);
    end

    outg=cat(1,outg,[ns,2^ns ,term1,term2]);

    if ns > 2 & abs(outg(ns,3)-outg(ns-1,3))/abs(outg(ns,3)) <= e_tol%   test for convergence of split iterations
     convflg = 1;
     nconv=ns;
    elseif ns==nmax
     nconv=ns+1
    end
end
 I_gn=outg(nconv,3);


' B                ns                npanel            I_g1                I_g2'
outg
% from function IG = Ig2b(a,b,nconv,n1,f) %level contours for solution surface  IG(np,ngauss) num panels np vs  Gauss
points n1
    %Num1 Simpson Gauss Scripts\Keep for Book\GaussMesh_wkg2_Book.m
%          Ig(a,b,np,n1,f)
%{a,b}; np panels ; %n1 =2,3,4,5,6,,9 point Gauss Quadr ;  f = function of x ns=[0:1:nconv]
%' Num1 Simpson Gauss Scripts\Keep for Book\GaussMesh_wkg2_Book.m '
nn1=[2:n1];
outsf= zeros(nconv,length(n1));
tmp0=[];
for j = nn1  %start loop over Gauss roots amd coef
m1=j-1;
for k=[0:1:nconv] %start loop over  multiple panels
  np=2^k;
  hp=(b-a)/np; %   h1k=hp/2 is initial xcale factor from f(x) to g(t)
  u=coef(1,m1).rc(1,:) ; %  u root values 'rn1i'

  if k==0            %start if for a separate single panel contribution
    x=(b-a)*u/2+(b+a)/2;
    cn1i=coef(1,m1).rc(2,:) ;
    % Gauss Quadra coeficients
      fi=f(x) ;
      gi=(hp/2)*fi ;
      tmp0=cat(1,tmp0,sum(cn1i.*gi) );
  elseif k > 0   %end if for 1 panel kk=0 and do multipanels as matrix
  % a,b to -1,1 Transf

  uu= repmat(u, np,1);
  kk=[1:np]';
  h2k= a+(2*kk-1)*hp/2 ;         % x [a,b]  to t {-1 ,1] xform ;  hp/2 is used to compute h1k
  kk=repmat(kk,1,j); % repmat to convert to  np x ngauss matrices
  x=a+(hp/2)*(uu+(2*kk-1)) ;
  cn1i=coef(1,m1).rc(2,:) ;
  cn1ii=repmat(cn1i,np,1);
  % Gauss Quadra coeficients
  fi=f(x) ;
```

```
      gi=(hp/2)*fi ;
      IG=sum(sum(cn1ii.*gi) );
      GKU= (hp/2)*f( ((2*kk-1)+uu )*hp/2 +a);
      CC=cn1ii;
      IG2=[sum(sum(CC.*GKU))];     % Dbl sum on Pt-by-Pt Matrix product "CC.*GKU"
      outsf(j,k)=IG2; %form IG2(Np, Ng) for trade-off
      end %end if for kk=1 and multiple panels
end     %end loop over multiple  panels
end    %end loop over Gauss roots amd coef


% ********** Plots and Tables ***************
nsplt=[0:1:nconv];
outsf2=reshape(outsf(find(outsf)),length(nn1),nconv);
outsf2=cat(2,tmp0,outsf2); % add in col for single panel
Imax=outsf2(end,end);
[xx,yy]=meshgrid(2.^nsplt,nn1) ; % meshgrid table points overlay on plots
if pltflg(1)==1
figure(1)
   if imag(Imax)==0
   surfc([2.^nsplt],[nn1],outsf2);colorbar ;hold on
   plot(xx,yy,'k*:');plot(xx,yy,'ro:')
   axis xy
   else
   surfc([2.^nsplt],[nn1],abs(outsf2));colorbar;hold on
   plot(xx,yy,'k*:');plot(xx,yy,'ro:')
   axis xy
   end
end


if pltflg(2)==1
 figure(2)
 el_max2=max(max(outsf2));
 el_min2=min(min(outsf2));
 delcontour2=(el_max2-el_min2)/20;
 V2=[el_min2:delcontour2:el_max2];
 [c2,s2]=contourf([2.^nsplt],[nn1],outsf2); hold on, shading flat;
 clabel(c2,s2,'fontsize',12,'color','r','rotation',0),colorbar
 plot(xx,yy,'k*:');plot(xx,yy,'ro:')
 nrow=size(outsf2,1);
 outsf2b=outsf2(nrow:-1:1,:);
 nn2=max(nn1):-1:2;
 nn2=[NaN,nn2];
 outsf_tbl=cat(2,nn2',cat(1,2.^nsplt,outsf2b))% ******  table of outputs ******
 Ntot=2.^(nsplt +1)-1;
 outsf3=nn1'*Ntot ;
 outsf3b=outsf3(nrow:-1:1,:);
 nn2=max(nn1):-1:2;
 nn2=[NaN,nn2];
 outsf_Feval=cat(2,nn2',cat(1,2.^nsplt,outsf3b))% ******  table of outputs ******
end

if pltflg(3)==1
 figure(3)
 axis xy
 [c3,s3]=contour([2.^nsplt],[nn1],outsf3,10);hold on %# contours for Num of function evaluations  over (ns, n1) plane
 clabel(c3,s3,'fontsize',12,'color','r','rotation',0)
 plot(xx,yy,'k*:');plot(xx,yy,'ro:')
end
```

```
if pltflg(4)==1
 figure(4)
 el_max4=max(max(outsf3));
 el_min4=min(min(outsf3));
 delcontour4=(el_max4-el_min4)/20;
V4=[el_min4:delcontour4:el_max4];
[c4,s4]=contour(2.^nsplt,nn1,outsf3,V4);hold on
clabel(c4,s4),colorbar
plot(xx,yy,'k*:');plot(xx,yy,'ro:')
end

if pltflg(7)==1
figure(7)
v=[Imax*.9:Imax*.05:Imax*1.1] ;
%[xx,yy]=meshgrid(2.^nsplt,nn1) ;
[c7,s7]=contour([2.^nsplt],[nn1],outsf2,v,'k-'); hold on; clabel(c7,s7)
plot(xx,yy,'k*:');plot(xx,yy,'ro:')
end

if pltflg(8)==1
figure(8)
[c8,s8]=contourf([2.^nsplt],[nn1],outsf2,v,'k-'); hold on;  clabel(c8,s8,'fontsize',12,'color','r','rotation',0),colorbar
plot(xx,yy,'k*:');plot(xx,yy,'ro:')
end
```

Explicit outputs for Ex. 8 { **'exp(3.*x).*sin(2*x.^2)'** } ; **[a, b]=[2, 5]** } and Ex. 9 { **'sin(50.*x)./(pi*x)'** ; [-1, 1.00001] }. The actual quadrature values I_G(Npanel, Ngauss) are given in a table and shown as black * with a surrounding red circle in the contour plots. The multi-panel Gauss expansions continue until the e_tol = 10^{-3} is met in each example, so clearly the most accurate stop value is at max #panels and max #Gauss points (10).   The contour MatLab® plots interpolate through the table points and give a visual of the (Np,Ng) regions in which the quadrature meets the e_tol criterion. A table of the number of functional evaluations, "Nfevals" is also given; contours of those tables are also output by the program but not shown here.

MatLab Scripts

'ex'	'eq'	'a'	'b'	'e_tol'	'nmax'	'nptGauss'	'I_exact'
8	'exp(3.*x).*sin(2*x.^2)'	2	5	0.0001	12	10	-160808.749439

MultiPanel 10-Pt Gauss Quadrature (2 Methods)

nsplits	npanels	I_g1	I_g2
0	1	132212.452809829	132212.452809829
1	2	-162809.521955145	-162809.521955145
2	4	-160808.779860362	-160808.779860363
3	8	-160808.748910035	-160808.748910035
4	16	-160808.749516421	-160808.749516422

Number_Fevals(Np,Ng)

Np/Ng	1	2	4	8	16
10	10	30	70	150	310
9	9	27	63	135	279
8	8	24	56	120	248
7	7	21	49	105	217
6	6	18	42	90	186
5	5	15	35	75	155
4	4	12	28	60	124
3	3	9	21	45	93
2	2	6	14	30	62

Gauss Quadrature I_G(Np, Ng)

Np/Ng	1	2	4	8	16
10	132212.452809829	-162809.521955145	-160808.779860363	-160808.748910035	-160808.749516422
9	307499.819787665	-146492.830611546	-160807.408742505	-160808.749453026	-160808.749438644
8	-512698.54228734	-200814.673706204	-160836.044241273	-160808.748373993	-160808.749438654
7	-843265.807834643	-193913.671654985	-160565.532089261	-160815.765004342	-160805.713882739
6	-361048.463084806	292645.565469789	-161425.743733655	-160808.359472745	-160808.749964483
5	-255769.49107611	-905343.834313842	-179257.900229285	-160762.968158614	-160808.699853576
4	956655.365440452	-325490.100599638	36027.9825746935	-163051.468907509	-160810.480989873
3	-516806.740379942	149569.253470416	-876528.345384199	-117515.318179186	-160884.467777225
2	305898.94197961	-158409.230023936	463456.623634732	-510808.480619138	-153773.721396114

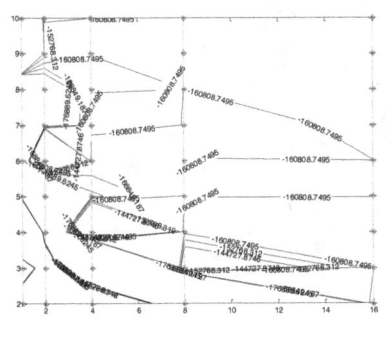

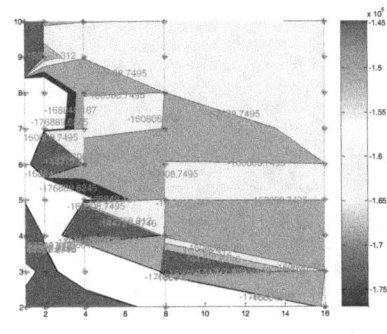

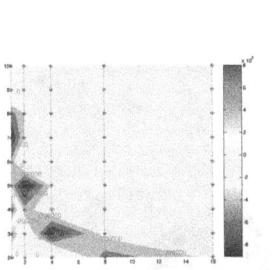

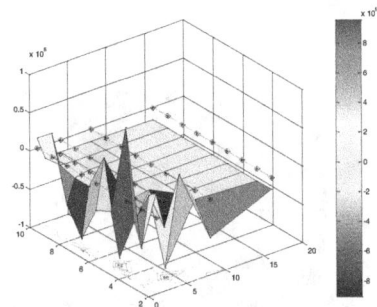

'ex'	'eq'	'a'	'b'	'e_tol'	'nmax'	'nptGauss'	'I_exact'
9	'sin(50.*x)./(pi*x)'	-1	1.00001	0.0001	12	10	1.000

MultiPanel 10-Pt Gauss Quadrature (2 Methods)

ns	npanel	I_g1	I_g2
0	1	1.28061539171328	1.28061539171328
1	2	1.44811695717567	1.44811695717567
2	4	0.987226479173282	0.987226479173281
3	8	0.987789269641588	0.987789269641587
4	16	0.98778927261578	0.987789272615782
5	32	0.987789273090309	0.987789273090308

Number_Fevals(Np,Ng)

Np/Ng	1	2	4	8	16	32
10	10	30	70	150	310	630
9	9	27	63	135	279	567
8	8	24	56	120	248	504
7	7	21	49	105	217	441
6	6	18	42	90	186	378
5	5	15	35	75	155	315
4	4	12	28	60	124	252
3	3	9	21	45	93	189
2	2	6	14	30	62	126

Gauss Quadrature I_G(Np, Ng)

Np/Ng	1	2	4	8	16	32
10	1.28061539171328	1.44811695717567	0.987226479173281	0.987789269641587	0.987789272615782	0.987789273090308
9	4.6115129827525	0.419521713823982	0.992283593768984	0.987789354550527	0.987789273094205	0.987789273094652
8	0.759034662801323	0.619620109568962	0.960979875835262	0.987786804760358	0.987789273150953	0.987789273094653
7	7.07805293021007	1.76125682390724	1.09783155449832	0.987906429278089	0.987801106938749	0.987789442949557
6	-0.338361300712786	1.98432827906333	0.73290384544479	0.986787730871223	0.987789685938249	0.987789273097685
5	9.76794122768957	0.632446810754807	1.07736909765188	0.999927306766001	0.987767226279691	0.987789272444177
4	-1.37913785088821	-0.62809477199806	1.87605673077116	0.894270377133963	0.988580123192834	0.987789367378125
3	14.5388308011762	-0.947558437924366	0.451474468002914	1.36332927063466	0.970583103192461	0.987780884247354
2	-0.616517914989751	-0.971408002914569	-1.07833730356731	0.725112645347707	1.17864984843705	0.988179643491871

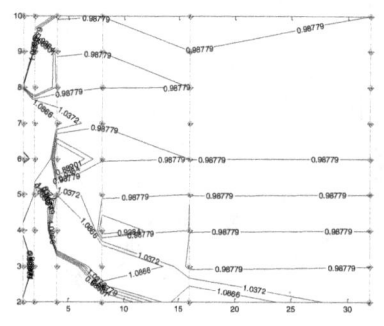
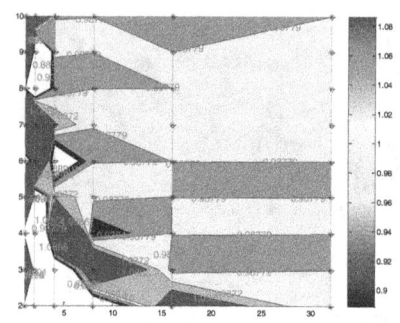

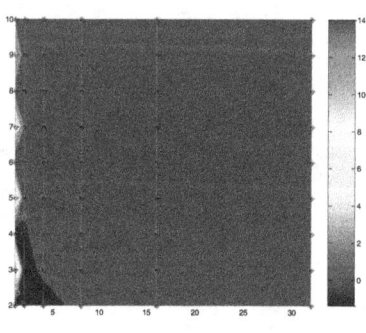
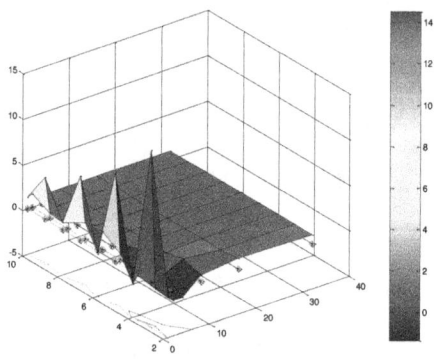

5.13 Simpson 1^d Composite with fixed intervals

```
%Fixed Composite Simpson
function In = I_n(a,b,np,eq)
%input fcn for lists with "." e.g.,  eq = 'x.^4'
f=inline(eq,'x');
h1=(b-a)/(2*np);
x_1=a+[1:2:2*np]*h1;
if np == 1
   In= (h1/3)*(f(a)+f(b)+4*f((a+b)/2));
elseif np >= 2
x_2=a+[2:2:2*np-1]*h1;
In= (h1/3)*(f(a)+f(b)+4*sum(f(x_1))+2*sum(f(x_2) ));
end
```

The following table compares the Adaptive Simpson quadrature method with the simple fixed Simpson quadrature given above. For the Ex. 7), the fixed Simpson algorithm appears to yield the same accuracy when evaluated for the same number of panels that the adaptive method stops at. However, for more difficult integrals such as Ex. 9) sin(50x)/(π x) the adaptive simpson method will perform better.

ex 7) '2*x.*cos(2*x)-(x-2).^2' ;a=0;b=5; e_tol = various

Adaptive Composite Simpson	Fixed Composite Simpson	
Itmp = -15.3063093510881 ; e_tol = 1e-005	64 panels	I_n(a,b,64,eq) = -15.3063079774007
Itmp = -15.3063079957494; e_tol = 1e-007	128 panels	I_n(a,b,128,eq) = -15.3063079851493
	256 panels	I_n(a,b,256,eq) = -15.3063079856205
Itmp = -15.3063079856533; e_tol = 1e-010	1024 panels	I_n(a,b,1024,eq) = -15.3063079856516
Itmp = -15.3063079856518; e_tol = 1e-012	2048 panels	I_n(a,b,2048,eq) = -15.3063079856517

5.14   Quadratures – Simple 2^d Simpson & Gaussian

```
% Simpson 2ᵈ Composite Quadrature for 2 x 2 Cell Region
%
clear all
format short
f=inline('x*y^2')
a=2.1; b=2.5;% x integration limits
c=1.2; d=1.4;% y integration limits
n1=2; n2=2; % 2 cells x 2 cells square integration region
h=(b-a)/(2*n1); k=(d-c)/(2*n2)
Iexact=.311573333333;
W=zeros(5);
fij=zeros(5);
i1=[0:2*n1];
j1=[0:2*n2];
```

$$F^T = \begin{array}{c|ccccc} & \begin{matrix}2.1\\x0\end{matrix} & \begin{matrix}2.2\\x1\end{matrix} & \begin{matrix}2.3\\x2\end{matrix} & \begin{matrix}2.4\\x3\end{matrix} & \begin{matrix}2.5\\x4\end{matrix} \\ \hline y0 & 3.0240 & 3.1680 & 3.3120 & 3.4560 & 3.6000 \\ y1 & 3.2813 & 3.4375 & 3.5937 & 3.7500 & 3.9063 \\ y2 & 3.5490 & 3.7180 & 3.8870 & 4.0560 & 4.2250 \\ y3 & 3.8272 & 4.0095 & 4.1917 & 4.3740 & 4.5562 \\ y4 & 4.1160 & 4.3120 & 4.5080 & 4.7040 & 4.9000 \end{array} \begin{matrix}1.20\\1.25\\1.30\\1.35\\1.40\end{matrix}$$

```
xi=a+h.*i1;
yj=c+k.*j1;
 fij=(yj.^2)'*xi;
```

$$W = \begin{array}{c|ccccc} & \begin{matrix}2.1\\x0\end{matrix} & \begin{matrix}2.2\\x1\end{matrix} & \begin{matrix}2.3\\x2\end{matrix} & \begin{matrix}2.4\\x3\end{matrix} & \begin{matrix}2.5\\x4\end{matrix} \\ \hline y0 & 1 & 4 & 2 & 4 & 1 \\ y1 & 4 & 16 & 8 & 16 & 4 \\ y2 & 2 & 8 & 4 & 8 & 2 \\ y3 & 4 & 16 & 8 & 16 & 4 \\ y4 & 1 & 4 & 2 & 4 & 1 \end{array} \begin{matrix}1.20\\1.25\\1.30\\1.35\\1.40\end{matrix}$$

```
W(1,:)=[1 4 2 4 1];
W(2,:)=[4 16 8 16 4];
W(3,:)=[2 8 4 8 2];
W(4,:)=[4 16 8 16 4];
W(5,:)=[1 4 2 4 1];
W
format long
Isimp = sum(sum((h*k/9)* (fij.* W)))
%   ****   Isimp = .3115733  ****
% Note that the transpose does not matter here because of the symmetry of both W and F
```

```
% Gaussian Quadrature 2ᵈ Example
clear all
%g =inline('.075*log(.3*u+.5*v+4.2)', 'u','v')
g =inline('.02*(2.3+.2*u)*(1.3+.1*v)^2', 'u','v')
u0=.77459; u1=0; u2=-.77459;
v0=.77459; v1=0; v2=-.77459;
c0=.55555; c1=.88889; c2=.55555;
G=[[g(u0,v0),g(u1,v0),g(u2,v0)]' [g(u0,v1),g(u1,v1),g(u2,v1)]' [g(u0,v2),g(u1,v2),g(u2,v2)]' ]
c=[c0,c1,c2]'
Q=c'*G*c
%Alternate
Cm=c*c'
Q2=sum(sum(Cm.*G))
```

MatLab Scripts

5.15 General 2^d Simpson Quadrature & Weight Matrix

%**Simp2d_Bookb_Arrayb** Num1\Num1 Book Scripts Current Quadrature Keep for Book
%2-Dim Simpson Quadrature with Non-Regular Cells curves c(x) and d(x)

```
% uses W=SimpWt2d_Book(n1,n2);
%Ex=1  I_ex=.03330556611;      fij=exp(yj/xi);                a=0.1; b=0.5 ;   c=xi.^3 ; d=xi.^2;
%Ex 2  I_ex =16.50864;         fij=( yj^3 + xi^2 );           a=2 ;  b=2.2 ;   c=xi;     d=2*xi;
%Ex 3  I_ex=1/6=.16666         fij=xi(i);                     a=0;   b=1 ;     c=xi ;    d=1;
%Ex 4  I_ex =8*pi/3 = 8.377580; fij(i,j)=2*pi*xi(i)^2;        a=0;   b=2 ;     c=xi ;    d=2;
%Ex 5  I_ex =8*pi/3  fij=4*sqrt(xi.^2+yj.^2)*(2-sqrt(xi.^2+yj.^2)); a=0; b=2 ; c=0 ; d=sqrt(4-x.^2);

clear all;  Ex=4;  n1=2; n2=2;

% ******* Cases ********
if Ex==1
a=0.1 ;  b=0.5 ;% Ex 1
elseif Ex ==2
a=2 ;  b=2.2 ; % Ex 2
elseif Ex ==3
a=0 ;  b=1;
elseif Ex ==4  %3d 45 deg cone in cylindrical r thet z coords  f = f(r)dr r*dthet*dz ; for r->x z->y ==>
2*pi*Int(Int(x*y,x,0,2),y,0,2-x)
a=0 ;  b=2;
elseif Ex ==5
 a=0;  b=2 ;
end


h=(b-a)/(2*n1);
W=zeros(2*n1+1,2*n2+1);
fij=zeros(2*n1+1,2*n2+1);% i(row) n1-cells  ; j(cols) n2-cells
yj=zeros(1,2*n2+1);
i1=[0:2*n1]; j1=[0:2*n2];
xi=a+h.*i1 ;   % Eval all xi coordinate as row vector x0,for each i:  i(row) n1 cells {a a+h a+2h ...a+2n1*h=b}

% ******* Cases ********
if Ex==1
c=xi.^3 ;  d=xi.^2; % Ex 1 integral limit evaluations c(x) & d(x)
elseif Ex==2
c=xi;  d=2*xi;     % Ex 2 integral limit evaluations c(x) & d(x)
elseif Ex ==3
 c=xi ;d=1 ;
elseif Ex ==4
 c=xi ; d=2 ;
elseif Ex ==5
 c=0 ;  d=sqrt(4-xi.^2);c=repmat(c,1,2*n1+1);
end
% *************
xitmp=repmat(xi',1,2*n2+1);
k=(d-c)/(2*n2);  % Eval all k values as row-vector over all j(col) subscripts
ktmp=repmat(k',1,2*n2+1);
ctmp=repmat(c',1,2*n2+1);
yjtmp=ctmp+k'*j1;

if Ex==1
    fij=exp(yjtmp./xitmp); % Ex 1
   elseif Ex==2
     fij=( yjtmp.^3 + xitmp.^2 );% Ex 2
   elseif Ex ==3
```

```
    fij =xitmp;
  elseif Ex ==4
    fij = 2*pi*xitmp.^2;
  elseif Ex ==5
    fij = 4*sqrt(xitmp.^2+yjtmp.^2).*(2-sqrt(xitmp.^2+yjtmp.^2));
  end
W=SimpWt2d_Book(n1,n2);
I_s=(h/9)* sum(sum((W.*fij').*ktmp'));
disp('Simp2d_Bookb_Arrayb.m')
disp('uses Wt Matrix: W = "SimpWt2d_Book(n1,n2)"' )
%format short g
format long g
'     Ex      n1      n2          I_s'
  [Ex  n1  n2   I_s]
```

% ******* 2^d Comparison of Gauss vs. Simpson ********

Ex 1 fij=exp(yj/xi); a=0.1; b=0.5; c=xi.^3; d=xi.^2

n	I_Gauss	I_Simpson	n
1	0.0333453874623914	0.03249434240621	2
3	0.0333058313348074	0.03330476252307	3
4	0.0333055670528154	*	
5	0.0333055661186752	0.03330546128190	5
10	0.0333055661145462	0.03330555954577	10
30	*	0.033305566035048	30

I_exact = 0.033305566110000

Ex2 fij=(yj^3 + xi^2); a=2; b=2.2; c=xi; d=2*xi

n	I_Gauss	I_Simpson	n
1		16.50865000000002	1
3	16.50864000000000	16.50864012345680	3
4	16.50864000000000	16.50864003906252	4
5	16.5086399999999	16.50864001600002	5
8	*	16.50864000244143	8
16	*	16.50864000015261	16
50	*	16.50864000000162	50

I_exact =16.50864000000000

Ex3 fij=xi;a=0;b=1;c=xi;d=1;

n	I_Gauss	I_Simpson	n
1	0.166666666666667	0.166666666666667	1
2	0.166666666666667	0.166666666666667	2
3	0.166666666666667	0.166666666666667	3

I_exact =1/6 = 0.16666666666667

Ex4 fij=xi; a=0; b=2; c=2-xi ; d=2; Cone Cyl. Coord.

n	I_Gauss	I_Simpson	n
1	8.37758040957278	8.37758040957278	1
2	8.37758040957278	8.37758040957278	2

I_exact =8*pi/3 = 8.37758040957000

%Simpson 2^d Weight Matrix

```
function y = SimpWt2d(n1,n2)
% Computes Simpson Wt Matrix for "n1 x n2" cells; %n1=3;n2=5;% yields r=7 c=11
%along X: h=(b-a)/2*n1;  % X: 3+2*(n1-1)= 2*n1+1  n1=3 yields cols = 7 for X
%along Y: k=(d-a)/2*n2 ; % Y: 3+2*(n2-1)= 2*n2+1  n2=5 yields rows = 11 for Y
 % Special Cases
smc0=[1 4 1];
smc1=[1 4 2];% 1st row of 1st cell
startc = smc1;
if n1 == 1 & n2 == 1    % single cell
   y=[1 4 1;4 16 4 ;1 4 1];
 elseif n1 == 2  & n2 == 1   % 2 cells along x , 1 cell along y
     y =[1 4 2 4 1;4 16 8 16 4 ;1 4 2 4 1];
 elseif n1 == 1  & n2 == 2    %1 cell along x , 2 cells along y (transpose ')
     y = [1 4 2 4 1;4 16 8 16 4 ;1 4 2 4 1]';
  elseif n1 == 2  & n2 == 2  % 2 cells along x , 2 cells along y
     y=[1 4 2 4 1;4 16 8 16 4 ;2 8 4 8 2;4 16 8 16 4 ;1 4 2 4 1];
  elseif ((n1 >= 2) | ( n2 >= 2))
smc2=[4 2];% to 1st row for two cells
mdlc = repmat(smc2,1,(n-2));
```

```
%add n1-2 "smc2"s for middle a%  mdlc = 4    2    4    2    4    2
finishc =[4 1];% finish n1 th cell
 fullc=[startc, mdlc,finishc]; %sm1 = [1 4 2 4 2 4 2 4 2 4 1]
sm1= fullc;
        %fullc = 1    4    2    4    2    4    2    4    2    4    1
sm2=4*sm1;  %sm2 = [4 16 8 16 8 16 8 16 8 16 4]
sm3=[sm1 ;sm2 ;2*sm1];
        % sm3 =
        %1    4    2    4    2    4    2    4    2    4    1
        %4    16    8    16    8    16    8    16    8    16    4
        %2    8    4    8    4    8    4    8    4    8    2
start=[sm1 ;sm2;2*sm1];
        %start =
        %1    4    2    4    2    4    2    4    2    4    1
        %4    16    8    16    8    16    8    16    8    16    4
        %2    8    4    8    4    8    4    8    4    8    2
mdl=[sm2; 2*sm1];
        %mdl =
        %4    16    8    16    8    16    8    16    8    16    4
        %2    8    4    8    4    8    4    8    4    8    2
mdl=repmat(mdl,n2-2, 1);% adjoin remaining n2-2 middle rows
        %mdl =
        % 4    16    8    16    8    16    8    16    8    16    4
        % 2    8    4    8    4    8    4    8    4    8    2
finish=[sm2;sm1];
        %finish =
        %4    16    8    16    8    16    8    16    8    16    4
        %1    4    2    4    2    4    2    4    2    4    1
   WMtrx=[start;mdl;finish];% Complete Simpson weight matrix "n1 x n2" cells
    y = WMtrx;
end
```

5.16 *General 3ᵈ Simpson Quadrature & Weight Matrix*

```
%  Simp3d_Book_Array.m
%          Uses Weight Matrix:    W = "SimpWt3d_Book.m"
%3-Dim Simpson Quadrature with Non-Regular Cells curves c(x) & d(x) and surfaces alp(x,y) & beta(x,y)

% *** run parameters: mflg= 0: total mass. M;  mflg= 1: 1st moment;  mflg= 2: 2nd moment

% *******  Examples Setup *******
% Ex=1: Full Cone:  I_ex = 8.37504476 ; fijk(xi,yj,zk2)= sqrt(xi^2 +yj^2) ;
% a=-2 ; b=2 ; c=-sqrt(4-xi(i)^2); d=sqrt(4-xi(i)^2);alp(xi,yj) = sqrt(xi(i)^2 +yj(j)^2) ; beta(xi,yj)=2;
% Ex=2 :  4*( 1/4-cone) multiply by 4 axial symmetry ; fijk(xi,yj,zk2)= sqrt(xi^2 +yj^2) ;
%a=0;  b=2 ;  c = 0;  d = sqrt(4-xi.^2);  alp(xi,yj) = sqrt(xi(i)^2 +yj(j)^2); beta(xi,yj)=2;
% Ex=3 : larger cone height =3   x in range  a=0 to b=+3

clear all
disp('Simp3d_Book_Array.m')
n1=20; n2=20; n3=20; Ex=1;  mflg=0;   % chooses z coord moments of fcn f(x,y,z)
     if mflg==0
         fijk=inline('zk.^0.*sqrt(xi.^2 + yj.^2)');    %*** total mass M. integration *****
     elseif mflg==1
         fijk=inline('zk.*sqrt(xi.^2 + yj.^2)');    %*** 1st  moment Mz of mass distribution  ***
     elseif mflg==2
         fijk=inline('zk.^2.*sqrt(xi.^2 + yj.^2)');    %*** 2nd moment Mz2 of mass distribution  ***
     end
%************* Initialize. *************
i1=[0:2*n1]; j1=[0:2*n2]; k1=[0:2*n3]; % index ranges for x,y,z steps

if Ex==1
  a=-2 ;  b=2 ;
elseif Ex==2
  a=0;  b=2 ;
  elseif Ex==3
  a=0;  b=3 ;
end

h=(b-a)/(2*n1);% scalar
xi=a+h.*i1 ;% Eval all xi  [a a+h a+2h ... a+2*n1*h=b]
xitmp=repmat(xi',1,2*n2+1);% (2*n1+1) x (2*n2+1) matrix

if Ex==1
  c = -sqrt(4-xi.^2);  d = sqrt(4-xi.^2);
elseif Ex==2
  c =0 ; c=repmat(c,1,2*n1+1);     %scalar c to constant row vector [ c c c ...]
  d = sqrt(4-xi.^2);
elseif Ex==3
  c = 0;  c=repmat(c,1,2*n1+1);     %scalar c to constant row vector [ c c c ...]
  d = sqrt(9-xi.^2);
end

k=(d-c)./(2*n2);   % Eval all k values as 1 x (2*n1+1) row-vector over all i subscripts
ktmp=repmat(k',1,2*n2+1);% 2*n1+1 x 2*n2+matrix
ctmp=repmat(c',1,2*n2+1);
yjtmp=ctmp+k'*j1; % 2*n1+1 x 2*n2+matrix
W=SimpWt3d_Book(n1,n2,n3);   %Weight Matrix dimensions  (2*n1+1) x (2*n2+1) x (2*n3+1)

    if Ex==3
```

```
        beta = 3 ;
    else
        beta = 2 ;
    end
alp = sqrt(xitmp.^2 +yjtmp.^2);% 2*n1+1 x 2*n2+matrix
 l =(beta-alp)./(2*n3);
temp=0;

 for kk=1:2*n3+1
    zk(:,:,kk)=alp+l*(kk-1);  % Eval  zk for all indices i, j, k2
    temp= temp+ W(:,:,kk).*fijk(xitmp,yjtmp,zk(:,:,kk)).*(ktmp.*l);
end
 if Ex==1
 I_s= h*sum(sum(temp))/27;
 else
 I_s=4*h*sum(sum(temp))/27;
 end
 format long g
'       Ex     n1      n2      n3      mflg     I_s '
          [Ex n1 n2 n3 mflg  I_s ]
```

% Simpson 3d Output Table for 45° Cone Moments

% Ex	n1	n2	n3	mflg	I_simpson
% 1	2	2	2	0	8.83377594169786
% 1	20	20	20	0	8.37833222375278
% 1	32	32	32	0	8.377779279234995
% 1	128	128	128	0	8.37758596063253
%					I_exact = 8*pi/3 = 8.37758040957278
% 1	20	20	20	1	13.4054146903771
%					I_exact = 64*pi/15 = 13.4041286553165
% 1	20	20	20	2	22.3426540959555
%					I_exact = 64*pi/9 = 22.3402144255274
% 2	2	2	2	0	8.44265651359811
% 2	3	3	3	0	8.39837596631408
% 2	5	5	5	0	8.38259185971219
% 2	10	10	10	0	8.37833222375277
% 2	20	20	20	0	8.37769761584039
% 2	32	32	32	0	8.37761427844768
% 2	50	50	50	0	8.3775909516632
% 2	128	128	128	0	8.37758133849012
%					I_exact = 8*pi/3 = 8.37758040957278
% 2	32	32	32	1	13.4041895286466
%					I_exact = 64*pi/15 = 13.4041286553165
% 2	32	32	32	2	22.3403320403143
%					I_exact = 64*pi/9 = 22.3402144255274
% 3	20	20	20	0	42.412094180192
% 3	128	128	128	0	42.4115055261062
%					I_exact = 27*pi/2 = 42.4115008234622
% 3	20	20	20	1	101.789171141278
%					I_exact = 162*pi/5 = 101.787601976309
% 3	20	20	20	2	254.473523778598
%					I_exact = 81*pi = 254.469004940773

%3^d Weight Matrix Function

```
function y = SimpWt3d_Book(n1,n2,n3)
% Computes Simpson Wt Matrix for "n1 x n2 x n3" cells
%*** 3d Special cases to be added for 3d generalization to cover all n1,n2,n3  ***
    if n1==1 & n2==1 & n3==1
      W3=zeros(3);
      W3(:,:,1) =   [1    4    1; 4    16    4; 1    4    1];
      W3(:,:,2) =   [4    16    4; 16    64    16; 4    16    4];
      W3(:,:,3) =   [1    4    1; 4    16    4; 1    4    1];
    y=W3;
   elseif n1==1 & n2==1 & n3==2
      W3=zeros(3,3,5);
      W3(:,:,1) =   [1    4    1; 4    16    4; 1    4    1];
      W3(:,:,2) =   [4    16    4; 16    64    16; 4    16    4];
      W3(:,:,3) =   [1    4    1; 4    16    4; 1    4    1];
      W3(:,:,4) = W3(:,:,2);
      W3(:,:,5) = W3(:,:,1);
    y=W3;

    elseif n1==1 & n2==2 & n3 == 1
      W3=zeros(3,5,3);
      W3(:,:,1) =[1 4 2 4 1;4 16 8 16 4 ;1 4 2 4 1];
      W3(:,:,2) =4*[1 4 2 4 1;4 16 8 16 4 ;1 4 2 4 1];
      W3(:,:,3) = W3(:,:,1);
    y=W3 ;
    elseif n1==2 & n2==1 & n3==1
      W3=zeros(5,3,3);
      W3(:,:,1) =[1 4 1; 4 16 4; 2 8 2;4 16 4;1 4 1];
      W3(:,:,2) =4*W3(:,:,1);
      W3(:,:,3) = W3(:,:,1)
    y=W3 ;

    elseif n1 == 2  & n2 ==2 & n3==1
      W3=zeros(5,5,3);
      W3(:,:,1) =[1 4 2 4 1;4 16 8 16 4 ;2 8 4 8 2;4 16 8 16 4;1 4 2 4 1];
      W3(:,:,2) =4*W3(:,:,1);
      W3(:,:,3) = W3(:,:,1);
    y=W3 ;

    elseif n1 == 1  & n2 == 2 & n3 == 2
      W3=zeros(3,5,5);
      W3(:,:,1) =[1 4 2 4 1;4 16 8 16 4 ;1 4 2 4 1];
      W3(:,:,2) =4*W3(:,:,1);
      W3(:,:,3) = 2*W3(:,:,1);
      W3(:,:,4) =4* W3(:,:,1);
      W3(:,:,5) = W3(:,:,1);
    y=W3 ;
   elseif n1 == 2  & n2 == 1 & n3 ==2
      W3=zeros(5,3,5);
      W3(:,:,1) =[1 4 1; 4 16 4; 2 8 2;4 16 4;1 4 1];
      W3(:,:,2) =4*W3(:,:,1);
      W3(:,:,3) = 2*W3(:,:,1);
      W3(:,:,4) =4* W3(:,:,1);
      W3(:,:,5) = W3(:,:,1);
    y=W3 ;
%*********** End of 3d Special cases ; Begin general case   *******************

    elseif ((n1 >= 2) | ( n2 >= 2) | n3 >= 2 )
```

MatLab Scripts

```
smc1=[1 4 2];% 1st three cols of row = of 1st cell plus "1" to join to next cell cols
startc = smc1;
smc2=[4 2];% join to 1st row for two cells
mdlc = repmat(smc2,1,(n2-2));%add n1-2 "smc2"s for middle  n2 is middle cols
finishc =[4 1];% finish n2_th cell in 1wst ro
                                      %fullc =[startc,        mdlc,        finishc]
fullc=[startc, mdlc,finishc];         %sm1 = [{1 4 2}   {4 2 4 2 4 2} {4 1}] all 2*n2+1 cols across 1st row
sm1= fullc;
sm2=4*sm1;                             %sm2 = [4 16 8 16 8 16 8 16 4] next row
sm3=[sm1 ;sm2 ;2*sm1];
start=[sm1 ;sm2;2*sm1];
mdl=[sm2; 2*sm1];
mdl=repmat(mdl,n1-2, 1);              % adjoin remaining m-2 middle rows
finish=[sm2;sm1];
WMtrx=[start;mdl;finish]; % Simpson WtMatrx "n1 x n2" cells 2*n1 +1 rows by 2*n2+1 cols


%***  Build in Z-direction with 2*n3+1 pages
        W3(:,:,1)= WMtrx;
        W3(:,:,2)=4*WMtrx;
        W3(:,:,3)=2*WMtrx;
        for i1=4:2:2*n3-2
                W3(:,:,i1)=4*WMtrx;
                W3(:,:,i1+1)=2*WMtrx;
        end
        W3(:,:,2*n3)=4*WMtrx ;
        W3(:,:,2*n3+1)=WMtrx;
        y=W3;
   end   %*** End of 3d general cases  ***
```

% single 3d cell output row col page								
% y(:,:,1) =			y(:,:,2) =			y(:,:,3) =		
% page#1			page#2			page#3		
% 1	4	1	4	16	4	1	4	1
% 4	16	4	16	64	16	4	16	4
% 1	4	1	4	16	4	1	4	1

5.17 General 2^d Gaussian Quadrature

%**Gauss2dQuadr_Book_b** Num1\Num1 Book Scripts Current\Keep for Book\Gauss2dQuadr_Book_b
%2-Dim Gauss Quadrature with Non-Regular Cells curves c(x) and d(x)

% uses Gauss Quadr tables for **2 to 7 and 10-point Quadratures**
clear all
Ex=2;
n1=10; n2=10;
m1=n1-1;
m2=n2-1;
coef(1,1).rc=[-1/sqrt(3) 1/sqrt(3); 1 1] ; %2 pt Gauss Quadr
coef(1,2).rc=[-sqrt(3/5) sqrt(3/5) 0 ; 5/9 5/9 8/9]; %3 pt Gauss Quadr
coef(1,3).rc=[-sqrt((15+2*sqrt(30))/35) sqrt((15+2*sqrt(30))/35) -sqrt((15-2*sqrt(30))/35) sqrt((15-2*sqrt(30))/35) ;
(18-sqrt(30))/36 (18-sqrt(30))/36 (18+sqrt(30))/36 (18+sqrt(30))/36];% 4 pt Gauss Quadr
coef(1,4).rc=[-0.5384693101056381 0.5384693101056381 -0.9061798459386640 0.9061798459386640 0.0 ;
0.4786286704993665 0.4786286704993665 0.2369268850561891 0.2369268850561891 .568888888888889];% 5 pt
Gauss Quadr
coef(1,5).rc=[-0.6612093864662645 0.6612093864662645 -0.2386191860831969 0.2386191860831969 -
0.9324695142031521 0.9324695142031521;
0.3607615730481386 0.3607615730481386 0.4679139345726910 0.4679139345726910 0.1713244923791704
0.1713244923791704];%6 pt Quadr
coef(1,6).rc=[-0.4058451513773972 0.4058451513773972 -0.74155311855993945 0.74155311855993945 -
0.9491079123427585 0.9491079123427585 0.0;
0.3818300505051189 0.3818300505051189 0.2797053914892766 0.2797053914892766 0.1294849661688697
0.1294849661688697 0.4179591836734694]; %7 pt
coef(1,9).rc=[-0.148874339 0.148874339 -0.433395394 0.433395394 -0.679409568 0.679409568 -0.865063367
0.865063367 -0.973906529 0.973906529;
0.295524225   0.295524225   0.269266719   0.269266719   0.219086363   0.219086363   0.149451349
       0.149451349   0.066671344   0.066671344]; %10 pt

%coef(1,4).rc=[LOOK UP]
%coef(1,2).rc
%ans = %-0.7746   0.7746   0
% 0.5556   0.5556   0.8889

%coef(1,2).rc(2,:)
%ans = 0.5556   0.5556   0.8889

%coef(1,3).rc
% -0.8611   0.8611   -0.3400   0.3400
% 0.3479   0.3479   0.6521   0.6521

%coef(1,3).rc(1,:)
% -0.8611   0.8611   -0.3400   0.3400

%Ex=1
%I_exact=.03330556611;
%fji(j,i)=exp(yj(j)/xi(i));
%a=0.1 ; b=0.5 ; c=xi.^3 ; d=xi.^2;

%Ex 2
%I_approx =16.50864
%fij(i,j)=(yj(j)^3 + xi(i)^2);
%a=2 ; b=2.2 ; c=xi; d=2*xi;

%Ex 3
%I_exact=.5;

```
%fij(ji,j)=xi(i);
%a=0;  b=1 ; c=xi ;  d=1;-


%Ex 4
%I_exact =8*pi/3 = 8.37758040957
%fij(i,j)=2*pi*xi^2;
%a=0;  b=2 ; c=xi ;  d=2;


%Ex 5
%I_exact =8*pi/3 = 8.37758040957   3d cartesian reduced to 2d cartesian
%fij(i,j)=4*sqrt(xi.^2+yj.^2)*(2-sqrt(xi.^2+yj.^2));
%a=0;  b=2 ; c=0 ;  d=sqrt(4-x.^2);


% ******* Examples ********
if Ex==1
a=0.1 ;  b=0.5 ;% Ex 1
elseif Ex ==2
a=2 ;  b=2.2 ; % Ex 2
elseif Ex ==3
a=0 ;  b=1;
elseif Ex ==4 | Ex==5 %3d 45 deg cone in cylindrical r thet z coords  f = f(r)dr r*dthet*dz ; for  r->x z->y  ==>
2*pi*Int(Int(x^2,x,0,2),y,x,2)
a=0 ;  b=2;
end


% a,b to -1,1 Transf
h1=(b-a)/2; h2=(b+a)/2 ;
u=coef(1,m1).rc(1,:); %u root values 'rn1i'
x=h1*u+h2;


% ******* Examples ********
if Ex==1
c=x.^3 ;  d=x.^2 ;% Ex 1 integral limit evaluations c(x) & d(x)
elseif Ex==2
c=x;  d=2*x;     % Ex 2 integral limit evaluations c(x) & d(x)
elseif Ex ==3
c=x;d=1 ;
elseif Ex ==4
c=x ; d=2 ;
elseif Ex ==5
c=0 ; d=sqrt(4-x.^2);
end


% *************
v=coef(1,m2).rc(1,:);  %v root values 'rn2i'
k1tmp=(d-c)/2; k1=repmat(k1tmp',1,n2);
k2tmp= (d+c)/2; k2=repmat(k2tmp',1,n2); %[k2' k2'.... k2'];
vtmp=repmat(v,n1,1);
yj=k1.*vtmp+k2 ;     % n1 x n2 matrix
xi=repmat(x',1,n2); % n1 x n2 matrix
%****************
cn1i=coef(1,m1).rc(2,:);
cn2j=coef(1,m2).rc(2,:);
Cij=cn1i'*cn2j; %2d Coeff Matrix n1 x n2 matrix
fij=zeros(n1,n2);% i(row) n1-pts ; j(cols) n2-pts % n1 x n2 matrix


% ******* Examples ********
if Ex==1
```

```
fij=exp(yj./xi); % Ex 1
elseif Ex==2
fij=yj.^3 + xi.^2 ;% Ex 2
elseif Ex ==3
fij =xi;
elseif Ex ==4
fij = 2*pi*xi.^2;
elseif Ex ==5
fij=4*sqrt(xi.^2+yj.^2).*(2-sqrt(xi.^2+yj.^2));
end
%   *************
end
end
gij=k1.*fij;
I_g=h1*sum(sum(Cij.*gij));

disp('Gauss2dQuadr_Book_b')


%format short g
format long g
'        Ex  n1  n2     I_g'
[Ex n1 n2  I_g]
```

% Simpson Composite *vs.* Gaussian Quadrature 2^d

Ex 1 fij=exp(yj/xi); a=0.1; b=0.5; c=xi^3; d=xi^2;

n	I_ Simpson	I_ Gauss	n
	*	0.0333453874623914	2
	*	0.0333058313348074	3
		0.0333055670528154	4
2	0.03249434240621	*	
3	0.03330476252307	*	
5	0.03330546128190	0.0333055661186752	5
10	0.03330555954577	*	
30	0.03330556603505	0.0333055661145462	10

I_exact = 0.03330556611000

Ex 2 fij=(yj^3 + xi^2); a=2; b=2.2; c=xi; d=2*xi;

n	I_ Simpson	I_ Gauss	n
1	16.50865000000002	*	
2	16.24616523437502	16.50864	3
4	16.50864003906252	16.50864	4
5	16.50864001600002	16.5086399999999	5
8	16.50864000244143	*	
16	16.50864000015261	*	
50	16.50864000000162	16.5086399990399	10

I_exact =16.50864000000

Ex 3 fij=xi; a=0; b=1; c=xi; d=1

n	I_ Simpson	I_ Gauss	n
1	0.16666666666667	*	
2	0.16666666666667	0.166666666666667	2
3	0.16666666666667	0.166666666666667	3

I_exact =1/6 = 0.166666666666667

Ex 4 fij=xi; a=0; b=2; c=2-xi; d=2;45°-Cone in Cyl. Coord.

n	I_ Simpson	I_ Gauss	n
1	8.37758040957278	8.37758040957278	2
2	8.37758040957278	8.37758040957278	3

I_exact = 8.37758040957278

Ex 5 fij=4*sqrt(xi^2+yj^2)*(2-sqrt(xi^2+yj^2)); a=0; b=2; c=0; d=sqrt(4-x^2);
3^d reduced to 2^d Cartesian

2	8.01804432288262
3	8.34054363148634
5	8.3750447446989
7	8.37703491194628
10	8.37751123138186

I$_{exact}$ =8*pi/3 = 8.37758040957

5.18 *General 3ᵈ Gaussian Quadrature*

```
%  Gauss3dQuadr_Book_Array.m Num1\Num1 Book Scripts Current\Keep for Book\Gauss3dQuadr_Book_Array
%3-Dim  Gauss Quadrature with Non-Regular Cells curves c(x) & d(x) and surfaces alp(x,y) & beta(x,y)


% run parameters: mflg= 0: total mass M;  mflg= 1 1st moment;  mflg= 2: 2nd moment
%  ******* Four Examples Setup *******
% *** Ex=1:    full cone ****** I_approx = 8.37504476 Burden & Faires
%fijk(xi,yj,zk2)= sqrt(xi(i)^2 +yj(j)^2) ;  %a=-2 ; b=2 ; c=-sqrt(4-xi(i)^2); d=sqrt(4-xi(i)^2);
% alp(xi,yj) = sqrt(xi(i)^2 +yj(j)^2) ; beta(xi,yj)=2; same for all cases
% *** Ex=2 : 4*( 1/4-cone) %a=0; b=2 ; %c = 0; d = sqrt(4-xi.^2); Use CYL Coord symmetry to solve Ex 1
% *** Ex=3 : 4*( 1/4-cone) a=0 to b=+3  c = 0; d = sqrt(9-xi.^2); larger cone height h=3
% *** Ex=4 : 4*( 1/4-cone) a=0 to b=+2; c = 0; d = sqrt(4-xi.^2);Same as Ex 2 REPLACE WITH NEW EX
clear all
disp('Gauss3dQuadr_Book_Array.m')
n1=5; n2=5; n3=5; Ex=1;  mflg=0;   % chooses z coord moments of fcn f(x,y,z)
m1=n1-1;m2=n2-1;m3=n3-1; %reduce index by 1 for Table lookup index
    if mflg==0
        fijk=inline('zk.^0.*sqrt(xi.^2 + yj.^2)');    %*** total mass M. integration *****


    elseif mflg==1
        fijk=inline('zk.*sqrt(xi.^2 + yj.^2)');   %*** 1st  moment Mz of mass distribution  ***
    elseif mflg==2
        fijk=inline('zk.^2.*sqrt(xi.^2 + yj.^2)');   %*** 2nd moment Mz2 of mass distribution  ***
    end
coef(1,1).rc=[ -1/sqrt(3)  1/sqrt(3); 1 1] ; %2 pt Quadr
coef(1,2).rc=[ -sqrt(3/5) sqrt(3/5) 0 ; 5/9  5/9 8/9]; %3 pt Quadr
coef(1,3).rc=[-sqrt((15+2*sqrt(30))/35) sqrt((15+2*sqrt(30))/35) -sqrt((15-2*sqrt(30))/35) sqrt((15-2*sqrt(30))/35);
(18-sqrt(30))/36  (18-sqrt(30))/36 (18+sqrt(30))/36 (18+sqrt(30))/36];%  4 pt Quadr
coef(1,4).rc=[-0.5384693101056381 0.53846953101056381 -0.9061798459386640  0.9061798459386640 0.0 ;
   0.4786286704993665 0.4786286704993665 0.2369268850561891 0.2369268850561891 .568888888888889];%5 pt
Quadr
coef(1,5).rc=[-0.6612093864662645 0.6612093864662645 -0.2386191860831969 0.2386191860831969 -
0.9324695142031521 0.9324695142031521;
   0.3607615730481386 0.3607615730481386 0.4679139345726910 0.4679139345726910 0.1713244923791704
0.1713244923791704 ];%6 pt Quadr
coef(1,6).rc=[-0.4058451513773972 0.4058451513773972 -0.74155311855993945 0.74155311855993945 -
0.9491079123427585 0.9491079123427585 0.0;
   0.3818300505051189  0.3818300505051189 0.2797053914892766 0.2797053914892766 0.1294849661688697
0.1294849661688697 0.4179591836734694 ]; %7 pt
coef(1,9).rc=[-0.148874339 0.148874339 -0.433395394 0.433395394 -0.679409568 0.679409568 -0.865063367
0.865063367 -0.973906529 0.973906529;
0.295524225    0.295524225    0.269266719    0.269266719    0.219086363    0.219086363     0.149451349
          0.149451349    0.066671344    0.066671344 ]; %10 pt


%************* Initialize. *************
Cijk=zeros(n1,n2,n3);  gijk=zeros(n1,n2,n3);
c1=coef(1,m1).rc(2,:);
c2=coef(1,m2).rc(2,:);
c3=coef(1,m3).rc(2,:);
u=coef(1,m1).rc(1,:);
v=coef(1,m2).rc(1,:);
w=coef(1,m3).rc(1,:);
if Ex==1
  a=-2 ; b=2 ;
elseif Ex==2 | Ex==4
  a=0;  b=2 ;
 elseif Ex==3
  a=0;  b=3 ;
end
```

```matlab
h1=(b-a)/2;
h2=(b+a)/2;
utmp=repmat(u',1,n2);
xi=h1*u+h2;  % u=coef(1,m1).rc(1,:); %u root n1=2 values 'rn1i'% n1=2 x-values
xitmp= h1*utmp+h2;% n1 x n2 array

   if Ex==1
   c = -sqrt(4-xi.^2);  d = sqrt(4-xi.^2);
   elseif Ex==2 | Ex ==4
   c =0 ; c=repmat(c,1,n1);      %scalar c to constant row vector [ c c c ...]
   d = sqrt(4-xi.^2);
   elseif Ex==3
   %c = -sqrt(9-xi.^2);  d = sqrt(9-xi.^2);
   c = 0;  d = sqrt(9-xi.^2);
   end

k1=(d-c)/2; k1tmp=repmat(k1',1,n2); %[k1'  k1'.... k1'];
k2=(d+c)/2; k2tmp=repmat(k2',1,n2); %[k2' k2'.... k2'];
vtmp=repmat(v,n1,1); %v root values 'rn2j'  %  n1 x n2 matrix ( 2 x 3 say)
yjtmp =k1tmp.*vtmp+k2tmp; %  n1 x n2 matrix %v n2=3 root values 'rn2i'% n1*n2=6 y-values
alp = sqrt(xitmp.^2 +yjtmp.^2);

   if Ex==3
       beta = 3 ; betatmp=repmat(beta,n1,n2);
   else
       beta = 2 ;betatmp=repmat(beta,n1,n2);
   end

l1=(betatmp-alp)/2;
l2=(betatmp+alp)/2;

   for kk=1:n3 % kk index so not to be confused with param k;  n3 = 4 say
   zk(:,:,kk)  = l1*w(kk)+l2;
   gijk(:,:,kk)=fijk(xitmp,yjtmp,zk(:,:,kk)).*(k1tmp.*l1); %n1 x n2 x n3 array 2x3x4
   Cijk(:,:,kk)=(c1'*c2)*c3(kk);  %n1 x n2 x n3 array
   end
CC=reshape(Cijk,n1*n2*n3,1);% long vector CC 2*3*4 = 24 elements
GG=reshape(gijk,n1*n2*n3,1);% long vector GG 2*3*4 = 24 "aligned" elements
Ig=h1*sum(CC.*GG);  %pt-by-pt mult; then just one sum; mult by h1
if Ex==2 | Ex ==3 | Ex==4
Ig=4*Ig; % 4*(1/4-Cone )
end
format long g
 ' Ex       n1       n2       n3        mflg     Ig  '
           [Ex n1 n2 n3 mflg  Ig]
```

3^d Gauss Quadrature Examples

Ex. 1 fijk=sqrt(xi^2 +yj^2); a=-2; b=2; c=-sqrt(4-xi^2); d=sqrt(4-xi^2); alp=sqrt(xi^2 +yj^2); beta =2

n_{Gauss}	$mflg$	I_{Gauss}	I_{Gauss} Cone
2	0	9.91817010071489	8.01804432288262
3	0	5.4599	8.34054363148634
4	0	8.74841	8.36946457238135
5	0	7.6729930985922	8.37504474469892
6	0	8.4999	8.37659718853037
7	0	8.1038	8.37703491194628
10	0	8.40641891694364	8.37751123138185
5	1	12.657	13.40038157064628
10	1	13.4334993013141	13.4040183655298
5	2	21.378	22.33346778697460
10	2	22.3780127469108	22.3400117122616

Ex=2 fijk=sqrt(xi^2 +yj^2);a=0; b=2; c = 0; d =sqrt(4-xi^2); Cyl. Coord; 4*(1/4-Cone)

n_{Gauss}	$mflg$	I_{Gauss}
2	0	8.01804432288262
3	0	8.34054363148634
4	0	8.36946457238135
5	0	8.37504474469892
6	0	8.37659718853038
7	0	8.37703491194626
10	0	8.37751123138185
5	1	13.4003815706463
10	1	13.4040183655298
5	2	22.3334677869746
10	2	22.3400117122616

Ex=3 fijk=sqrt(xi^2 +yj^2); a=0; b=3; c=0; d =sqrt(9-xi^2) *Large Cone*

n_{Gauss}	$mflg$	I_{Gauss}
2	0	40.5913493845933
3	0	42.2240021343996
4	0	42.3704143976806
5	0	42.3986640200383
5	1	101.759147552095
5	2	254.392156511008

6 References

1. *"Scientific Computing"* , Heath, M. , McGraw-Hill, Boston, 2002.
2. *"Numerical Analysis, 9h Ed."* Burden, R. L., Faires,J.D., Brooks-Cole, CA, 2011.
3. *"Introduction* to *Scientific Computing,"* Van Loan, C.F., Prentice-Hall, NJ, 2000.
4. *"Numerical Methods Using MatLab 4th Ed.,"* Mathews,J.H., Fink, K.D., C.F., Prentice-Hall, NJ, 2004.
5. *"Numerical Recipes in Fortran 77,2nd Ed.,"* Press, W.H.,Teukolsky, S.A., Vetterling, W.T., Flannery, B.P.,Cambridge, UK, 1996.
6. *"Numerical Analysis - A Practical Approach,2nd Ed.,"* Maron, M.J., Macmillan Publishing Co., New York, 1987.
7. "Numerical Methods and Fortran Case Studies," Dorn, W.S., McCracken, D.D., John Wiley & Sons, New York, 1972.

7 Index

A

accelerating convergence of iterates, 2-30, 2-34
    example, 2-32, 2-33
accumulation point, 1-18
accuracy, 1-2
accuracy tolerance, 2-2
adaptive quadrature, 4-18
    flow diagram, 4-19
    motivation, 4-18
adding new data nodes, 3-4, 4-29, 4-40
Aitken's Δ^2 method, 2-30
    derivation, 2-34
    example, 2-32, 2-33
    iterate difference ratios, 2-31
    MatLab script, 5-5
algorithm error, 1-12, 1-13
algorithms behaving badly, 1-3, 1-4
    numerical derivative, 1-5
    polynomial cancellation effects, 1-8
    standard deviation, 1-10
allocated tolerances, 4-19

B

backward step derivative formula, 4-5
backward Δ-polynomial, 3-24, 3-25
base, 1-14
basis polynomials, 3-4, 3-12, 3-26
Bessel function
    iterated interpolation, 3-19
    short table, 3-2
Bezier curves, 3-36, 3-37, 5-11
    equations, 3-35
    interactive, 5-12
    parametric polynomials, 3-7
binomial coefficient, 3-25
bisection method, 2-4, 2-14, 2-20, 2-21
boundary conditions, 3-29
bracketing methods, 2-3

C

calculus theorems, 1-17, 1-21, 1-22, 1-23
Cauchy method, 2-12, 2-14
change of variable, 4-31
Chebyshev method, 2-21, 2-23
Chebyshev root finding method, 2-22
closed Newton-Cotes, 4-14
coefficients of interpolating polynomial, 3-20
composite 2^d Simpson, 4-35, 4-36
    non-rectangular cell, 4-38
composite 3^d Simpson
    non-rectangular cell, 4-39
composite methods, 4-15
composite Simpson method, 4-15, 4-34
composite trapezoidal method, 4-20
    recursion formula, 4-20
computational algorithm, 1-2
computational efficiency, 2-28, 2-29, 3-4
computational load, 4-29

computationally expensive, 3-26
computer graphics, 3-33, 3-37
computer word, 1-14
computer word size, 1-25
condition number, 1-28
constant step size, 3-24, 3-25
continuity, 1-20
    at a point, 1-18
    function defined on sequence, 1-19
convergence, 2-11, 3-4
    of integral, 4-32
    of sequence, 3-19
    rate of, 1-2
coordinate transformations, 3-24, 3-25
    adaptive coordinate systems, 3-3
    Gauss quadrature mapping, 4-29
    inversion, 4-30, 4-31
cross multiplying, 3-18
cubic spline, 3-29
    algorithm, 3-29, 3-30
    equations for coefficients, 3-28
    polynomial, 3-7, 3-28, 3-30
    sub-splines, 3-30

D

derivatives, 4-1, 4-2
    3 and 5 point formulas, 4-4
    best choice formula, 4-5
Derivatives & Integrals, 4-1
difference equations, 1-30, 1-31
difference of iterates, 2-2
differentiable function:, 1-20
differential equations, 3-3
diminishing returns, 4-17
divergent integrands, 4-30
divergent solution, 1-30, 1-32
divided differences, 3-21, 3-29
    constant stepsize Δ-table, 3-25
    DD table, 3-33
    example, 3-22
    example inverse interpolation, 3-23
    polynomial coefficients, 3-20
double precision, 1-33, 1-37
dynamic range, 1-2, 1-14, 1-15

E

economization procedure, 3-31, 3-32
efficient numerical computation, 3-14
end point singularities
    example, 4-33
error analysis, 1-4, 1-24
error growth, 1-30
error tolerance, 2-17, 3-15
evenly spaced nodes, 4-29
exact quadrature, 4-23
excised point, 1-18, 1-20
experimental accuracy error, 1-12
experimental validation, 1-13
exponential growth, 1-30, 1-32
extrapolation
    higher accuracy 1st derivative estimate, 4-2, 4-12

Index

notional table, 4-8
Romberg integration, 4-20, 4-21

F

false position method, 2-5, 2-14
finite digit arithmetic, 3-21
first derivative, 4-3, 4-5
    3-point formula, 4-6
fixed point method, 2-15, 2-16, 2-20, 2-21
    examples, 2-19
    map the box region, 2-17, 2-18
    MatLab script, 5-4
    trajectories, 2-18
fixed point theorem, 2-17
floating point representation, 1-3, 1-14, 1-26
    base 10 example, 1-15
    typical computer systems, 1-16
forward step derivative formula, 4-5
forward Δ-polynomial, 3-24, 3-25
free and clamped, 3-29, 3-30
function, 1-18, 1-20
functional decomposition, 2-15
functional evaluations, 4-29, 4-35
functional iteration, 2-15, 2-16, 2-17
functional value, 2-2

G

Gauss root finding method, 2-22, 2-23
Gaussian Non-Rectangular Cells – 3^d, 4-41
Gaussian quadrature, 4-23
    2-point, 3-point, 4-24
    3-point, 4-40
    comparison to composite methods, 4-29
    Gauss-Legendre table, 4-24
    intuitive proof, 4-25
    multiple integrals, 4-34, 4-40
    n-point, 4-26
    other orthonormal polynomials, 4-27
Gauss-Legendre quadrature, 4-28
Gauss-Legendre table, 4-28
Gauss-Tschebyshev quadrature, 4-28
Gauss-Tschebyshev table, 4-28
general solution to difference equation, 1-31
generate polynomials on the fly, 4-26
geometric series, 2-34
guide points, 3-33, 3-34, 3-35

H

Halley method, 2-12
Hermite polynomials, 3-5, 3-6, 3-26, 3-27, 3-33, 4-27
    divided difference tableau, 3-26
    forward, backward coefficients, 3-27
    properties, 3-8
Horner's method, 2-24, 2-25, 2-26, 2-29, 3-17
    computational efficiency, 2-28
    example, 2-27
h-structure, 4-8
hybrid root finding methods, 2-3

I

improper integrals, 3-3, 4-1, 4-30

end point singularities, 4-32
    examples, 4-31, 4-33
    Simpson estimate, 4-33
improved iterate, 2-30
infinite limits, 4-30
infinite sum, 2-34
initial conditions for difference equation, 1-31
instability, 4-7
integrals, 4-1, 4-2
    convergent, 4-32
    integrating by parts, 4-30
    other Gaussian quadratures, 4-27
integration cells for 2^d integral, 4-34
integration methods comparison, 4-29
interior points, 3-29
intermediate value theorem, 1-22, 2-1
interpolating polynomials, 3-2, 4-1, 4-2
interval [a,b], 4-13
Introductory Material, 1-1
inverse Newton interpolation polynomials, 3-23
invertible function, 1-18
iterated Lagrange interpolation, 3-17
    example, 3-19
    tableau, 3-18

J

Jacobian partial derivatives matrix, 2-13

L

Lagrange basis polynomials, 3-12, 3-13, 3-18
    example, 3-14
Lagrange polynomial expansion, 3-5, 3-6, 3-11, 3-12, 4-2, 4-3, 4-4, 4-6
    integral of, 4-13
    interpolation table, 3-15
    iterated interpolation, 3-17
    truncation error, 3-16
Laguerre polynomials, 4-27
    properties, 3-8
least squares fit, 3-2
Legendre polynomials
    first four, 4-26
    properties, 3-8
limit of sequence, 1-19
linear combination, 1-31, 3-4
linear convergence, 2-11, 2-30
linear interpolation, 3-11, 3-15, 3-16
linear Lagrange expansion, 3-14
logarithmic divergence, 2-2
loss of significant digits, 1-11

M

machine epsilon, 1-14, 4-21
machine precision, 4-21
mantissa, 1-14
mathematical instability, 1-28
MatLab Scripts 5-1
    Aitken's Convergence Acceleration, 5-5
    Bezier Curves, 5-11
    Bezier Curves - Interactive, 5-12
    Derivative of exp(x), 5-2
    Fixed Point Example, 5-4

Index

General 2^d Gaussian Quadrature, 5-24
General 2^d Simpson Quadrature & Weight Matrix, 5-17
General 3^d Gaussian Quadrature, 5-27
General 3^d Simpson Quadrature & Weight Matrix, 5-20
Lagrange Polynomials for Sin(x) and Asin(x), 5-7
Polynomial Cancellation, 5-3
Quadratures – Simple 2^d Simpson & Gaussian, 5-16
Richardson Extrapolation, 5-10
Root Finding Methods Comparison, 5-13
Taylor Polynomials for 1/x, 5-6
matrix equation for cubic splines, 3-30
maximum degree of exactness, 4-23
maximum error for table, 3-15
mean value theorem, 1-21, 1-22
modeling error, 1-12
monomials, 3-31, 3-32
Muller's root finding method, 2-8
comparison with other methods, 2-20, 2-21, 2-22, 2-23
rate of convergence, 2-14
relation to secant method, 2-7
multiple integrals, 4-34, 4-35, 4-36, 4-40
Gauss functional evaluation pattern, 4-40
Simpson multi-cell weight matrix, 4-36
Simpson single cell weight matrix, 4-35
Multiple Integrals Methods Comparison, 4-43

N

n+1 point derivative formula, 4-4
nesting of polynomials, 1-9, 1-35, 2-24, 2-28, 3-14
Horner's method, 2-25
Newton polynomials
coefficients, 3-20
divided difference table, 3-20
forward, backward, central, 3-21, 3-22, 3-23
Newton-Cotes methods, 4-14, 4-30
truncation error, 4-17
Newton-Raphson root finding method, 2-9
convergence, 2-14
convergence issues, 2-10
example, 2-20, 2-21, 2-22, 2-23
generalization to higher degree, 2-12
Horner's method, 2-26, 2-27
polynomial roots, 2-29
root multiplicity & convergence, 2-11
two dimensions, 2-13
nodes
2^d nodal grid, 4-23, 4-34, 4-35
Hermite polynomial, 3-26
nodal derivatives, 3-26
nodal interval, 3-16, 3-18, 3-30, 4-29, 4-34
nodal points, 3-7, 3-13, 4-1, 4-2
nodal table, 3-14, 3-18, 3-20
nodal values, 3-2, 3-12, 3-26
non-bracketing methods, 2-3
non-convergent integral, 4-33
number conversion error, 1-12
number of free conditions, 4-23
number of iterations, 2-17, 2-19
number of points, 4-23
number of steps N, 3-15
numerical derivative, 4-7
comparison plot 2,5,8,10 s.d., 1-7
iterates to zero, 1-6, 5-2
plot and table, 1-6
Richardson extrapolation, 5-10

numerical instability, 1-30, 1-31
numerical quadrature, 4-1, 4-2, 4-13
numerical solution, 1-2
numerically efficient computing, 2-25, 4-21, 4-22, 4-36

O

object in free fall, 3-25
open Newton-Cotes, 4-14
optimal point, 4-8
orthonormal, 3-3
integration quadratures, 4-27, 4-28
polynomials, 4-27, 4-30
osculating polynomial, 3-7

P

parametric curves, 3-33
parasitic solutions, 1-30, 1-32
pattern for 2^d quadratures, 4-34
polynomial, 1-35
basis, 3-13
cancellation effects, 5-3
deflation, 2-29
degree, 4-23
finding all roots, 2-29
interpolation, 3-1
oscillations, 3-17
root finding, 2-24
Polynomial Interpolation, 3-1
position and velocity data, 3-4, 3-26
precision, 1-2, 1-14, 1-31, 1-32, 4-12, 4-21
proper integral, 4-30

Q

quadratic coefficients, 2-8
quadratic convergence, 2-11
quadratic formula, 1-34
quadratic interpolation, 3-11
quadratic Lagrange expansion, 3-14
quotient, 2-24

R

rates of convergence, 2-1, 2-14
ratio
of consecutive iterate differences, 2-30, 2-31
of consecutive iterate errors, 2-30, 2-31
ratio function for Newton-Raphson, 2-11
rationalize, 1-34
recursion relation, 2-24, 2-25, 2-26, 3-31, 4-21, 4-26
relative error, 1-14, 2-2
remainder, 2-24, 2-26
Richardson extrapolation, 4-2, 4-7, 4-8, 4-18
derive 5 point derivative, 4-10
math details, 4-9
stopping criterion, 4-11, 4-12
table, 4-22
robustness, 1-2, 2-14, 3-4
Rolle theorem, 1-21
generalized, 1-23
Romberg integration, 4-20
example, 4-21, 4-22
root finding techniques, 2-2, 2-3, 2-14

Index

comparison table, 2-20
comparison table, 2-20
comparison table, 2-21
comparison table, 5-13
non-bracketing comparison table, 2-22, 2-23
Root Finding Techniques, 2-1
roots, 1-31
    of polynomial, 2-24, 2-29
round off error, 1-25, 1-26, 1-33, 4-1, 4-7, 4-8, 4-12
    constant, 4-17
    tolerable, 4-7
round off instability, 4-17

S

scale parameters, 3-34
scientific computing, 1-2
    anatomy of an error, 1-12
    environment, 1-13
secant line, 2-9
    sequence, 1-20
secant root finding method, 2-6
    cf. Newton-Raphson, 2-9
    comparison table, 2-20, 2-21, 2-22, 2-23
    convergence rate, 2-14
    Muller generalization, 2-7
secant-tangent theorem, 1-21
significant digits, 1-4, 3-19
Simpson Gauss Quadratures Comparison - Cone Example, 4-42
Simpson quadrature, 4-14, 4-15
    2-panel example, 4-33
    coefficient pattern, 4-15
    comparison table, 4-29
    composite n/2-panels, 4-16
    truncation order, 4-17
singular points, 1-28, 4-30
singularity, 4-31
special polynomials, 3-4
stability, 1-2, 1-32, 4-1
    composite integral methods, 4-17
    numerical derivative, 4-17
stepsize, 1-26, 3-15, 4-7, 4-8
    fixed, 3-24
stopping criterion, 2-2, 2-4, 2-5, 2-6, 3-17, 4-11
structure of function, 1-20
sub-spline coefficients, 3-28
subtractive cancellation, 1-33, 1-35, 1-37
    reduce effects of, 1-36
sufficient conditions, 2-17
sum over 2^d cells, 4-34
symmetric derivative formula, 4-5
synthetic division, 2-24, 2-26, 2-27

T

table
    Bessel function, 3-2
    integration methods comparison, 4-29
    lookup with specified error, 3-15

    of nodal values, 4-5
tableau, 2-26, 2-27, 3-17
    Aitken's improved iterate, 2-30, 2-32, 2-33
Taylor polynomials, 1-25, 1-36, 2-9, 3-5, 3-6, 3-32, 4-6, 4-32, 4-33
    example, 3-10
    expansion, 2-12
    for 1/x, 5-6
Taylor's theorem, 1-23, 3-9
total error, 1-27
trade off, 3-15, 4-7
transformations, 4-27
transformed integral, 4-24
trapezoidal quadrature method, 4-14
    comparison table, 4-29
    composite n-panel, 4-16
    relation to Romberg quadrature
        1-panel and 2-panel, 4-21
trial solution, 1-31
tri-diagonal matrix, 3-29
trigonometric function, 3-31
trigonometric substitution, 4-31
truncation error, 1-25, 1-26, 3-9, 3-10, 3-16, 3-32, 4-1, 4-7, 4-8, 4-12, 4-22, 4-32
    composite 2^d Simpson, 4-35
    estimate, 4-3, 4-19
    known h dependence, 4-9
Tschebyshev method, 2-12, 2-20
Tschebyshev polynomials, 3-7, 3-31, 4-27
    economization, 4-28
    example, 3-32
    properties, 3-8
two-dimensional Taylor series, 2-13

U

unique fixed point, 2-17, 2-18
unstable solution, 1-32
upper bound, 3-9, 3-16
    truncation error, 4-13

V

Vandermonde polynomial, 3-5, 3-6
variation in the integrand, 4-18

W

Weierstrass Approximation Theorem, 3-3
weighted sum, 4-13, 4-23
weighting function, 3-8, 4-27, 4-28, 4-30

Z

zero truncation error, 4-21
zeros of Legendre polynomial, 4-26
z-notation, 3-27